Storey's Guide to Raising Dairy Goats

STOREY'S GUIDE TO
RAISING
DAIRY GOATS

FIFTH EDITION

BREED SELECTION • FEEDING • FENCING
HEALTH CARE • DAIRYING • MARKETING

Jerry Belanger & Sara Thomson Bredesen

Storey Publishing

The mission of Storey Publishing is to serve our customers by
publishing practical information that encourages
personal independence in harmony with the environment.

EDITED BY Deborah Burns and Sarah Guare
ART DIRECTION AND BOOK DESIGN BY Jeff Stiefel
TEXT PRODUCTION BY Liseann Karandisecky
INDEXED BY Samantha Miller

COVER PHOTOGRAPHY BY © Jason Houston, front and © Joan Budai/123RF, back and spine
INTERIOR PHOTOGRAPHY CREDITS appear on page 272.
ILLUSTRATIONS © Elara Tanguy, based on original artwork by Elayne Sears
DIAGRAMS pages 19, 21, 59 left, 87, 117, 182 and 195–199 by Ilona Sherratt

Storey Publishing
210 MASS MoCA Way
North Adams, MA 01247
storey.com

Printed in China by R.R. Donnelley
10 9 8 7 6 5 4 3

Library of Congress Cataloging-in-Publication Data

Names: Belanger, Jerome D., author. | Thomson Bredesen, Sara, author.
Title: Storey's guide to raising dairy goats / Jerry Belanger and Sara
    Thomson Bredesen.
Other titles: Guide to raising dairy goats
Description: 5th edition. | North Adams, MA : Storey Publishing, 2018. |
    Includes bibliographical references and index.
Identifiers: LCCN 2017034221 (print) | LCCN 2017047859 (ebook) |
    ISBN 9781612129334 (ebook) |
    ISBN 9781612129358 (hardcover : alk. paper) |
    ISBN 9781612129327 (pbk. : alk. paper)
Subjects: LCSH: Goat farming. | Goats.
Classification: LCC SF383 (ebook) | LCC SF383 .B44 2018 (print) | DDC
    636.3/9—dc23
LC record available at https://lccn.loc.gov/2017034221

CONTENTS

1

BASIC INFORMATION ABOUT GOATS

Perhaps your heart has been captured by the antics of baby goats at play, or your self-sufficient spirit has lead you on a course toward finding a family-friendly source of fresh milk, homemade cheese, and maybe some nontraditional meat products. Whatever the inspiration, this book assumes that you are interested in goats but does not assume that you know anything about them. A good place to start the journey is the very beginning, with some basic terms and facts.

Terms to Know

Female goats are called does or sometimes, if they're less than a year old, doelings. Males are bucks, or bucklings. Young goats are kids. In polite dairy goat company, they are never "nannies" or "billies," although you might hear these terms applied to meat goats. Correct terminology is important to those who are working to improve the image of the dairy goat. People who think of a "nanny goat" as a stupid and smelly beast that produces small amounts of vile milk will at least have to stop to consider the truth if she's called a doe instead.

Goat Myths and Truths

Over the many centuries and generations that goats have been humankind's companions and useful domesticated stock, myths have been passed along that have their origins in goat behavior and characteristics. As myths tend to be, however, these are exaggerated truths or downright fiction.

The Truth about Goat Aroma

Does are not smelly, they are not mean, and of course they don't eat tin cans. They are dainty, fastidious about where they walk and what they eat, intelligent (smarter than dogs, some scientists tell us), friendly, and a great deal of fun to have around.

Bucks have two major scent glands located between and just to the rear of the horns or horn knobs and minor ones in the neck region. Bucks do smell, but the does think it's great, and some goat raisers don't mind it either. The odor is strongest during the breeding season, which usually runs from September to about January. The scent glands can be removed, although some authorities frown on the practice because a descented buck can be less efficient

Not only can dairy goats provide milk and meat for a family, but their curiosity and typically gentle demeanor make them great companions as well.

at detecting and stimulating estrus and will still have enough of an odor to be mildly offensive.

Still, even if they don't stink, bucks have habits that make them less than ideal family pets. For instance, they urinate all over their front legs and beards or faces. This is natural, but it tends to turn some people off.

In most cases the home dairy won't even have a buck (see chapter 9), so you can keep goats even if you have neighbors or if your barn is fairly close to the house, and no one will be overpowered by goat aroma.

Livestock or Pets?

One of the challenges of goat public relations is that everyone seems to have had a goat in the past or knows someone who did. Most of them were pets, and that's where the trouble lies.

A goat is not much bigger than a large dog (average weight for a doe is less than 150 pounds [68 kg]), is no harder to handle, and does make a good pet. But a goat is not a dog. People who treat it like one are asking for trouble, and when they get rid of the poor beast in disgust, they bring trouble down on all goats and all goat lovers. If the goat "eats" the clothes off the line or nips off the rose-bushes or the pine trees, strips the bark off young fruit trees, jumps on cars, butts people, or tries to climb in a lap when it is no longer

> ### DON'T TIE HER DOWN
>
> Never stake out a goat. There is too much danger of strangulation, and many goats have been injured or killed by dogs. A goat that is tied can neither defend itself adequately nor escape to high ground. Even the family pet you thought was a friend of the goat could turn on it.

a cute little kid, it's not the goat's fault but the owner's.

Goats are livestock. Would you let a cow or a pig roam free and then damn the whole species when one got into trouble? Would you condemn all dogs if one is vicious because it was chained, beaten, and teased? Children can have fun playing with goats, but when they "teach" a young kid to butt people and that kid grows up to be a 200-pound (90 kg) male who still wants to play, there's bound to be trouble. Likewise, a mistreated animal of any species isn't likely to have a docile disposition.

Because goats are livestock, and more specifically dairy animals, they must be treated as such. That means not only proper housing and feed but also strict attention to and regularity of care. If you can't or won't want to milk at

> ### GOAT HOUSING
>
> Like a cow or a pig, a goat needs a sturdily fenced pen. Each doe requires a minimum of 10 square feet (0.9 sq m) of inside space, plus as much outdoor space as you can manage. Goats do not require pasture, and unless it contains browse, they probably won't utilize much of it anyway; they'll trample more than they eat. It's better to bring their food to them and feed them in a properly constructed manger, especially in a land- and labor-intensive small-farm situation. Managed intensive grazing is catching on, even in relatively small properties. For more on that option, see chapter 6.

12-hour intervals, even when you're tired or under the weather; if the thought of staying home weekends and vacations depresses you and you can't count on the help of a friend or neighbor, then don't even consider raising goats. The rewards of goat raising are great and varied, but you don't get rewards without working for them.

Goats Eat Everything, Don't They?

The goat (*Capra hircus*) is related to the deer — not to dogs, cats, or even cows. It is a browser rather than a grazer, which means it would rather reach up than down for food. The goat also craves variety. Couple all that with its natural curiosity, and nothing is safe from at least a trial taste. Lacking fingers, goats use their lips and tongues to investigate their world like an infant stuck in the oral stage. Anything hanging, like clothes on a wash line, is just too much for a goat's natural instincts to resist.

Rosebushes and pine trees are high in vitamin C, and goats love them. Leaves, branches, and the bark of young trees are a natural part of the goat's diet in the wild. If you expect them to mind their manners when faced with the chance of a garden smorgasbord, of course you'll have problems! But don't blame the goat.

Goats are not lawn mowers. Most of them won't eat lawn grass, unless starved to it, and they won't produce much milk on it.

Goats eat tin cans? Of course not. But they'll eat (or at least taste) the paper and glue on tin cans, which is what probably started the myth. Goats can be raised in a relatively small area. If there are no zoning regulations restricting livestock, dairy goats can be (and are) raised even on average-size lots in town.

A Little History

Goats have been humanity's companions and benefactors throughout recorded history — and even before. There is evidence that goats were among the first, some say *the* first, animals to be domesticated by humans, perhaps as long as 10,000 years ago. They provided meat, milk, skins, and undoubtedly entertainment and companionship.

Humans have had a long partnership with goats that can be traced back to petroglyphs found on rock walls in Iran.

Wild goats originated in Persia and Asia Minor (*Capra aegagrus*), the Mediterranean basin (*Capra prisca*), and the Himalayas (*Capra falconeri*). There were domesticated goats (*Capra hircus*) in Switzerland by the middle period of the Stone Age, and the first livestock registry in the world was organized in Switzerland in the 1600s — for goats.

Early explorers and voyagers distributed goats around the world, often carrying them on board ships as a source of milk and meat. There were goats, for example, aboard the *Mayflower* on its famous voyage to America in 1620, and British explorer Captain James Cook has become infamous in New Zealand and other South Pacific islands for dropping goats on dry land along his route. They were supposed to be emergency food in the event of subsequent shipwrecks.

As a consequence, goats spread to most parts of the world and ended up on shores far from home. Many returned to their feral state in their new homes, but today many more are in domesticated settings. The Food and Agriculture Organization of the United Nations (FAO) estimates there are nearly 994 million goats in the world, and the number is increasing at a rate of 2.4 percent a year.

In Europe, goats provided more milk than cows did until well after the Middle Ages. With the growth of modern cow dairies in densely populated countries, it is hard to say where goat milk consumption stands in the world today, but the FAO lists China as first in goat milk production and India as second. Goats are certainly more common in less fertile, more arid, or developing countries than they are in the United States and Canada, because they're more efficient animals than cattle in their ability to convert plants into more valuable animal protein. Although goats are more labor intensive than cattle, this is of small concern in backyard dairies and nonindustrialized countries, and of no concern at all where there isn't enough feed for cattle to do well or where a cow would produce more milk than a family could use.

THE BIOLOGY OF GOATS

Goats are mammals, of the class Mammalia: their young are born alive and suckle on a secretion from the mammary glands, which of course is milk.

They are of the order Artiodactyla, which means they are even-toed, hoofed mammals, and of the suborder Ruminantia (from the Latin, meaning "to chew cud") and have four "stomachs" like cows.

They belong to the family Bovidae, which among other things means that they have hollow horns that they don't shed. (Some goats are naturally hornless, or polled. Many more are disbudded: the horn buds are burned out with a hot iron or with caustic before they start to grow. Some goats are dehorned: the horns grow but are then cut off. See chapter 7.)

Goats belong to the genus *Capra*, which includes only goats. We will discuss the species *Capra hircus*, the domestic goat; within the species, subdivisions are known as breeds.

Breeds of Goats

While all domestic goats have descended from a common parentage, there are many breeds, or subdivisions of the species, throughout the world—more than 80. Only a few of these are found in the United States.

Goat breeds are classified according to their main purpose: that is, meat, mohair, or milk. In this book, we concentrate on the goats that have been bred for milk production, although in most respects care is the same for all.

Bear in mind that many, perhaps most, American goats are not purebreds: they are mixed and can't be identified as belonging to any particular breed. If these are fairly decent animals, they're usually referred to as "grades"; if not, most people call them "scrubs."

French Alpine

The French Alpine originated in the Alps and arrived in the United States in 1920, imported by Dr. C. P. DeLangle. The color of Alpines varies greatly and can range from solid shades to a variety of patterns. Often one animal displays several colors and shades. Plain white and the fawn and white markings of a Toggenburg are discriminated against.

There are recognized color patterns, such as the **cou blanc** (French for "white neck"). This goat has a white neck and shoulders, which shade gradually through silver gray to a glossy black on the hindquarters, and gray or black head markings. Another color pattern, the **chamoisée**, can be tan, red, bay, or brown, with black markings on the head, a black stripe

French Alpines have erect ears, and many of them have distinctive color patterns.

down the back, and black stripes on the hind legs. The **sundgau** has black and white markings on the face and underside. The **pied** is spotted or mottled; the **cou clair** has tan to white front quarters shading to gray, with black hindquarters; and the **cou noir** has black front quarters and white hindquarters.

According to the American Dairy Goat Association, Alpines average 2,548 pounds (1,156 kg) of milk a year, with 3.2 percent butterfat. The record is 6,990 pounds (3,171 kg).

You might also hear of British Alpines, Rock Alpines (named not because they like to climb on rocks any more than other goats do but because they were developed in America by Mary Edna Rock), and Swiss Alpines.

The stereotypical Alpine is pushy in a herd setting, will do anything for food, and is a little hyperactive. As with all stereotypes, this is a broad generalization, and there are many that don't fit that picture.

LaMancha

The LaMancha is a distinctly American breed. There's no mistaking a LaMancha: it looks as though it has no ears!

During the 1930s, Eula F. Frey of Oregon crossed some short-eared goats of unknown

---

MEASURING UP

Milk is generally measured in pounds with 8 pounds equal to 1 gallon (3.5 kg = 3.75 liters). A goat's height is measured to the withers, which is the sharp ridge of bone where the top of the neck joins the back.

LaManchas are noted for their "lack" of ears and are claimed by some to be the most docile breed. They also tend to be good milk producers.

The Nigerian Dwarf takes less space than full-size breeds and is an excellent choice for small farms.

origin with her top line of Swiss and Nubian bucks. The result was the LaMancha.

If you show LaManchas at the county fair, you'll have to put up with many exclamations of "What happened to the ears!?" Some people who are somewhat more knowledgeable about livestock will accuse you of allowing the animals' ears to freeze off. Even worse, you might be accused of cutting them off. But you don't milk the ears, LaMancha backers say. These goats have excellent dairy temperament, and they're very productive. A good average is 2,323 pounds (1,054 kg) of milk, with 3.7 percent butterfat.

If LaManchas have a personality quirk, it is that they tend to be the uncontested herd queens when put in with other breeds. One way goats create a pecking order is by nipping ears, so LaManchas can sit back and watch the others jostle for position. Although they can hear perfectly well, they are like teenagers — they play deaf when it suits them.

Nigerian Dwarf

Although these miniature dairy goats have been considered more of a novelty than true dairy animals for many years, the American Dairy Goat Association officially recognized this breed for their registry in 2005. Introduced in the early 1980s, when they were seen mostly in zoos, some of these little imports are excellent milkers for their size. As more serious breeders continue to develop them, their milk production is constantly increasing. What's more, they are considered dual-purpose animals, providing both milk and meat. Consequently, this breed is of particular interest to the backyard or small farmer.

The Nigerian Dwarf was the breed chosen for the Biosphere 2 experiment, in which eight people spent 2 years (from 1991 to 1993) sealed inside a self-contained, mostly self-sufficient dome in Arizona, along with 3,500 plant and animal species and no outside supplies or support except electricity. Biosphere 2 was designed as a space-colony model, though ecological research became the primary, scientific goal. At any rate, future space travelers might be milking Nigerian Dwarfs!

One Nigerian Dwarf doe gave a whopping 6.3 pounds (2.9 kg) of milk on test day, and another had 11.3 percent butterfat. A well-bred and well-managed Nigerian can be expected to produce an average of a quart (1 L) a day over a 305-day lactation. Many of these good producers have teats as large as those of the full-size breeds and are milked just as easily.

Nigerian Dwarf conformation is similar to that of the larger dairy breeds. All parts of the body are in balanced proportion. The nose is straight, ears are upright, and any color or combination is acceptable. Does can be no more than 22½ inches (57 cm) tall; bucks no more than 23½ inches (60 cm) tall. Weight should be about 75 pounds (35 kg). Being oversize for the breed standard is a disqualification in a goat show, as are a curly coat, a Roman nose, pendulous ears, and evidence of myotonia (a muscle condition characteristic of "fainting" goats).

Nigerian Dwarfs offer several advantages to the home dairy. Three Dwarfs can be kept in the space needed by one standard goat, so with staggered breedings a year-round milk supply is easier to achieve. This is enhanced by the Dwarf's propensity to breed year-round (compare this with seasonal breeding, discussed in chapter 10). These small goats can be kept on places that might not have room for larger

animals. Also, for some people, a regular goat will produce too much milk, while the Dwarf's quart-or-so a day is just fine. And the smaller animal is obviously easier to handle and transport, an attribute that many folks find especially appealing.

One potential disadvantage: many people still regard Nigerian Dwarfs as pets. If you purchase one from someone other than a dairy breeder, chances are the goat does not come from a line that has been upgraded and bred for milk production. She may not give enough milk to make it worth a trip to the barn, and if she has never been bred, she may have physiological problems that prohibit her from being bred in the future. Animals like this are not ideal choices for the home dairy.

Nubian

The most popular pure breed in America is the Nubian. Nubians can be any color or color pattern, but they're easily recognized by their long drooping ears and Roman noses. Unfortunately for people who like peace and quiet, that nose acts like the bell of a horn. Nubians are noted for loud voices, a tendency to stubbornness, and an unqualified dislike of rain, but the babies are so darned cute it's easy to overlook the personality flaws.

It's commonly said that the Nubian originated in Africa, but technically, the genealogy is a bit more complicated. From Africa, the Nubian made a stop along the way in its journey to the United States. Our Nubians are descendants of the Anglo-Nubian, which resulted from crossing native English goats with lop-eared breeds from Africa and India. The first three Nubians arrived in this country in 1909, imported by Dr. R. J. Gregg of Lakeside, California. The thicker-bodied

African genetics still show up in many herds in the United States. People looking for a dual-purpose animal that will maximize meat production probably want the thicker neck, shoulders, and loin, but those wanting higher milk production will prefer the more refined and angular variety.

The Nubian is often compared with the Jersey of the cow world. The average Nubian produces less milk than the average goat of any other breed, but the average butterfat content is higher. This is a good breed for cheesemakers; not so good for dieters.

Averages can be misleading, though. While the average production for a purebred Nubian is about 1,920 pounds (871 kg) of milk in 305 days with 93 pounds (42 kg) or 4.8 percent of butterfat, the top Nubian recorded by the American Dairy Goat Association produced 6,416 pounds (2,910 kg) of milk and 309 pounds (140 kg) of butterfat in 305 days. That's 802 gallons (3,036 L) of milk in 10 months.

Nubians are readily identified by their pendulous ears and Roman noses and should not be confused with the much shorter-legged Boer meat goat, which traditionally has a distinctive brown head, white body, and very bulky frame.

Oberhasli

There are no more Swiss Alpines. No, they're not extinct. In 1978, their name was changed to Oberhasli (oh-ber-HAAS-lee). This goat was developed near Bern, Switzerland, where it is known as the Oberhasli-Brienzer, among other names.

The outstanding feature in the appearance of the Oberhasli is its rich red bay coat with black "trim." The black includes stripes down the face, ears, back, belly, and udder. The legs are also black below the knees and hocks. Oberhasli milk production averages 2,004 pounds (909 kg) of milk, with 3.7 percent butterfat. The record is 4,665 pounds (2,116) of milk in 305 days.

Because Oberhaslis are fairly new to U.S. breed records, some people feel the purebred gene pool is a little shallow. Breeders will argue the point, but keep in mind that it may be more difficult to find unrelated breeding stock close to home when it's time to think about babies.

The Oberhasli is a rich bay color with black stripes on the face, ears, belly, udder, lower legs, and back. The American Dairy Goat Association allows for all-black coats but only on does.

Saanens are always light cream or white and have "dished" or concave faces.

Saanen

Next in popularity behind the Nubian is the Saanen (pronounced SAH-nen). This is a light cream or pure white goat with erect ears and a "dished" face that is just the opposite of the Nubian's. Saanens originated in the Saane Valley of Switzerland and have enjoyed a wider distribution throughout the world than any other breed. The first Saanens arrived in the United States in 1904.

They are large goats, with high average milk production: almost 2,740 pounds (1,243 kg) in 305 days. Butterfat averages 3.2 percent on a yearly basis. The all-time milk record is 6,571 pounds (2,980 kg).

Until recently, Saanens that were not pure white or light cream were discriminated against in purebred circles. Any that were colored or spotted could not be registered, and they were frequently disposed of. That changed in the 1980s when some Saanen breeders kept the colored or patterned animals, found that they were fine dairy animals, and started promoting them as a separate breed. They're not crossbreds; they're actually purebred Saanens but with a "color defect" that results when both the sire and the dam carry a recessive color gene. Today these goats are called Sables and were recently accepted as a separate breed by the American Dairy Goat Association. Since they are essentially "Saanens in party clothes," they won't be described separately here.

Saanen owners like to describe the personality of their breed as "laid back." A commercial producer with hundreds of animals would probably jump at that trait, but for the family dairy, it might not be as important.

Toggenburg

Toggenburgs are the oldest registered breed of any animal in the world, with a herdbook that was established in Switzerland in the 1600s. They were the first imported purebreds to arrive in the United States, in 1893, and have always been popular. Poet Carl Sandburg had a well-known herd of Toggenburgs.

Toggs, as they're sometimes affectionately called, are always some shade of brown with a white or light stripe down each side of the face, white on either side of the tail on the rump, and white below the hocks and knees.

Toggenburgs produce an average of 2,283 pounds (1,036 kg) of milk, with 3 percent butterfat. That's a little short of what the Saanens and Alpines average, but a Toggenburg currently holds the all-time record for milk production from a dairy goat, with 7,965 pounds (3,613 kg) — an astounding 995 gallons (3,767 L) of milk a year, from one little goat!

Someone walking into a herd of look-alike Toggenburgs will wonder how the owner can tell one from another, but each has its own personality and color modification that can be recognized once the eyes adjust.

Toggenburgs have white markings on the face, lower legs, and rump.

Other Breeds and Uses

New breeds are still being created. (Mating a doe of one breed to a buck of another produces a crossbred; creating a new breed is much more involved than that and generally takes years.) Recently created breeds include the Kinder (a Pygmy/Nubian cross), the Pygora (Pygmy/Angora), and the Santa Theresa (another dual-purpose breed). Although these breeds have very enthusiastic, usually regional, backers, they are rare compared with the eight recognized breeds, and most are still in the early stages of development.

You might hear about a few other rare breeds, such as the Tennessee Fainting Goat or Wooden Leg, which goes by several other names as well. When startled by a loud noise, the goat's muscles contract, and it tips over in what looks like a faint. Formerly considered a curiosity, Fainting Goats are becoming more popular in meat goat herds.

African Pygmy

Another dwarf breed gaining in popularity is the African Pygmy, often referred to simply as the Pygmy. This breed was first seen in the United States in the 1950s, and then only in zoos. These little goats are only 16 to 23 inches

> ### IT'S IN THE GENES
>
> In goat genetics, white is dominant and black is recessive. The white color pattern on Toggenburgs, including the vertical white stripes on both sides of the face, is dominant.

(40 to 58 cm) tall at the withers at maturity, and does weigh only 55 pounds (25 kg). They are very cobby (stocky, compact, and well muscled) — quite unlike a standard dairy animal. They look much like a beer keg with legs.

Despite their tiny size, some Pygmies are said to produce as much as 4 pounds (1.75 kg) of milk a day — that's half a gallon (almost 2 L) — and 600 to 700 pounds (270 to 320 kg) a year. And while the lactation period is shorter than for full-size goats (4 to 6 months rather than 10 months), the butterfat content often exceeds 6 percent. They are not really considered to be milk goats, and there is the practical problem of getting even a small bucket under them.

The Pygmy is more likely than the other breeds to have triplets or even quadruplets. They are registered by the National Pygmy Goat Association.

> ### HOW MANY GOATS?
>
> The 2016 agricultural census recorded 375,000 dairy goats in the United States, up 3 percent from a year earlier, but since then, at least two states have added one or more dairies with more than 5,000 animals each. The leading state by far is Wisconsin, followed by California, Iowa, and Texas. However, the census covers only farmers (that is, those meeting minimum income requirements), and most homestead and backyard animals aren't included in these totals.
>
> Who has the least? Some states, such as Hawaii, Rhode Island, and Delaware, have such small goat numbers that the statisticians lump them in with other states.

Some African Pygmies can produce enough milk for a small household, but they are generally considered pets.

Angora goats are raised primarily for mohair and meat.

Angora

Raised primarily for their long silken mohair, Angora goats have become quite popular in recent years. They are also raised for meat. While the fiber aspects are beyond the scope of this book, basic feeding, breeding, and management are similar for both Angora and dairy goats, except that you do not remove the horns on an Angora goat. The horns act as a cooling system for the hair-covered bodies.

Working Goats

Goats have proved useful as working animals, too. Wethers, or neutered males, are commonly used for packing, and goats of any breed can be trained to pull a small cart or wagon. In our technologically driven society, it's also good to see that goats are being put to practical use as "brushers" along roadways, power lines, and other places where weeds and brush need to be cleaned out. In many parts of the country, leasing out a herd of brush goats can be a lucrative business. It's also good for the environment.

Meat Goats

And then there are meat goats—animals raised for that specific purpose. The demand for goat meat has grown tremendously in recent years, due largely to ethnic markets (see chapter 15). Spanish and Angora goats were the traditional meat animals in the United States, but the Boer, originating in South Africa, has become hugely popular among meat goat ranchers. When they were first imported via New Zealand in 1993, a breeding buck could fetch as much as $70,000. We have since come to our senses and expanded herd numbers to the point where top prices are significantly less than that. I've never heard of anyone milking a Boer or other meat goat, but dairy goats provide plenty of meat as a by-product when culls and unwanted kids are butchered.

Selecting a Breed

Which breed is best? There is no answer to that question. If your reason for raising goats is to have a home milk supply, a goat that produces 1,500 pounds (680 kg) of milk a year is as good as any other goat, regardless of breed, that produces a like amount. You might not need or want a purebred at all, at least at first. Mixed-breed goats are much easier to find, are usually cheaper, and in some cases, produce more milk than purebreds.

Even for those interested in purebred stock, the choice of a breed isn't made because of any breed superiority or rational factors. If you are intending to make prodigious amounts of cheese, you might be influenced by breeds that tend to produce higher butterfat, but there is no guarantee that you will also get the higher protein levels that yield lots of cheese. On a small scale, the little bit of extra solids will hardly show in the finished product. You can, however, taste more butterfat in your milk, and some say it is creamier and sweeter . . . and also fatter.

In most cases, a breeder just "likes the looks" of a particular breed. It's also easier to find certain breeds than others, because the popularity of each varies from place to place. You might get a certain doe just because she's available, but if there is also a convenient stud service and a more likely market for her kids, you'd be making a wise choice.

So You Want a Goat?

This brief look at some of the basic facts about goats should help you decide if you really want to raise goats. I hope you do—but with full awareness of what will be expected of you. That means you'll want a lot more information on care and management. By all means, be sure to read that section of the book before you go looking for your goat, but even before we get to that, let's take a closer look at the product that probably led you to goats in the first place: milk.

PEDIGREE OR NOT?

A herdbook is the livestock version of a family tree and lists every registered animal back to the breed's "official" beginning. It certainly isn't necessary to buy a pedigreed animal to enjoy the benefits of fresh milk and entertainment, but there are some compelling reasons to consider it in Chapter 3.

2

MILK

One of the first questions a prospective goat owner who is interested in a family milk supply asks is, "How much milk does a goat give?" While the question is logical and valid, it's something like asking how many bushels of corn an acre of land can produce. How good the soil is, how much fertilizer is applied, what variety of seed is planted, how much of a problem weeds and insects are, and the amount of heat and moisture at the proper stages of development are all factors that affect the outcome. The same can be said of how much milk a goat produces. It depends.

How Much Milk?

There can be no set answer to the question of how much milk a goat will give, but here are some considerations.

Lactation Curves

It must first be understood that all mammals have lactation curves that, in the natural state, match the needs of their young. Humankind has altered these somewhat through selection to meet human needs, but they're still there.

The supply of milk normally rises quite rapidly after parturition (kidding, or freshening, or giving birth) in response to the demands of the rapidly growing young. In the goat, the peak is commonly reached about 2 months after kidding. From the peak, the lactation curve gradually slopes downward as the kid begins eating forage and gradually weans away from an all-milk diet.

This brings up what is probably the most common problem with terminology in reference to production: we often hear of a "gallon milker." The term has little or no practical value, because we want to know at what point in the lactation curve this gallon-day occurred and, even more importantly, what the rest of the curve looks like. The goat that produces a gallon a day 2 months after kidding and then drops off drastically and dries up a short time later will probably produce much less milk in a year than the animal whose peak day is less spectacular but who maintains a fairly high level over a long lactation. Especially in the home dairy, where a regular milk supply is the goal, slow and steady is more desirable than the flashy one-day wonder.

In addition, a "gallon" (3.75 L) is neither an accurate nor a convenient unit of measure for milk. Milk foams, and what if a goat gives just over or under a gallon? A gallon and 1 cup is tough to measure and even tougher to record. It's much more practical to speak of pounds of milk per lactation. As mentioned in the last chapter, a gallon weighs very close to 8 pounds

Dairy goats are fun to raise, but there are added benefits that come from being able to drink your own fresh, wholesome goat milk.

(3.5 kg). The traditional lactation period is 305 days. If a goat is to be bred once a year and dried off for 2 months before kidding for rest and rehabilitation, this period is logical.

The average 305-day lactation period is a convenient way to compare animals (cows are judged in similar fashion), but it is mainly for record purposes. The backyard goat dairy has no need to adhere to such a schedule, and in practice even most commercial dairies milk an animal for shorter or longer periods, depending on the animal's production. In some cases, it might not be worth dirtying the milk pail for a quart or so. In others, even a cup of milk might be considered valuable.

Tools and equipment have changed over time, but milk weight has long been a valuable measure of a family goat's worth.

Actually, many household goat dairies with animals that exhibit long lactations would do well to milk them for as long as they can without rebreeding. Production could be lower the second year, but this would be offset by avoiding a 2-month layoff, breeding expenses, and unwanted kids — including the considerable amount of milk kids will drink if they are not disposed of at birth. It should be pointed out, however, that not many goats will milk for that long: most will be dry before the 10 months are out (see chapter 10).

Average Production Levels

Looking at averages can be meaningless — after all, how many American families really have 2.4 children? — but sometimes that's the only way to get even a rough idea of a situation. Just remember that when a breed averages 2,000 pounds (900 kg) of milk, some of the goats are producing 3,000 pounds (1,350 kg), and some are milking only 1,000 pounds (450 kg). Some may even be producing less than that.

TRACKING LACTATION

The 305-day lactation period is simply an average; most goats milk for more or less than 305 days. According to the U.S. Department of Agriculture (USDA), only one-third of all does with official Dairy Herd Improvement Association (DHIA) records milked for 305 days. The milk production of many does declines sharply with the onset of seasonal estrus, or heat periods; after estrus, the does are dried off.

THE LACTATION CURVE

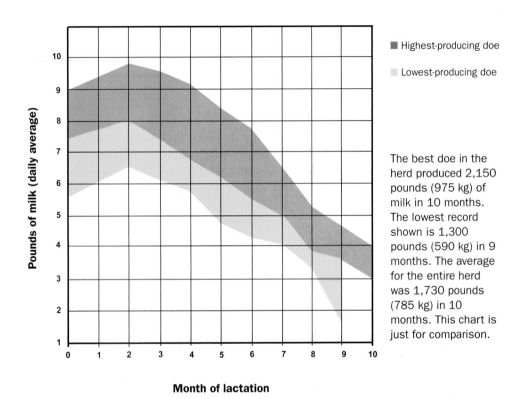

Highest-producing doe

Lowest-producing doe

The best doe in the herd produced 2,150 pounds (975 kg) of milk in 10 months. The lowest record shown is 1,300 pounds (590 kg) in 9 months. The average for the entire herd was 1,730 pounds (785 kg) in 10 months. This chart is just for comparison.

For many years, we said a decent average was 1,500 pounds (680 kg) a year, but that average is increasing. In 1998, for instance, the average for all large dairy breeds was 1,794 pounds (815 kg). In 2007, it was 2,241 pounds (1,015 kg), and in 2015, it was 2,352 pounds (1067 kg). But then the question arises, is it increasing because more good goats are on test while less productive animals aren't? Fifteen hundred pounds — about 187 gallons (708 L) — is still a reasonable expectation for the beginning goat farmer. But bear in mind that this average of 187 gallons over a period of 305 days doesn't mean you can plan on 0.6 gallon (2.25 L) a day. Remember the lactation curve.

Breed Records

Breed records are even more meaningless for the home dairy than are averages. The new goat owner has about as much chance of even coming close to record production as the guitar-pickin' kid down the road has of coming up with a hit song. It takes knowledge, experience, work, and maybe even a few lucky breaks, to produce a winner in any field.

At least the records will show you what goats are capable of. And they also demonstrate what a bucket of worms you get into when you ask, "How much milk does a goat give?"

You won't start out with a record setter, and you hope you won't get stuck with an underproducer, but it would be nice to find one that's "average." The only way to know for certain how much milk a goat gives is to milk her, weigh the milk, and record it for the entire lactation period. Or purchase a goat from someone who's been doing that.

Using Production Records

Leaving the pacesetters for a moment, let's look at the lactation curve on page 19 again. These are actual production records of a small herd of Nubians. The top doe produced 2,150 pounds (975 kg) in 10 months, and the bottom

SAMPLE BREED RECORDS

The Nubian breed record of 6,416 pounds (2,910 kg) of milk is well above the average of 1,795 pounds (815 kg), and many Nubians milk well below that number. The Saanen breed record was set in 1997 with 6,571 pounds (2,980 kg) of milk, but the 1998 average was only 1,899 pounds (860 kg), or 4,672 pounds (2,120 kg) less. The all-time all-breed U.S. production record was set by a Toggenburg in 1997: 7,965 pounds (3,615 kg).

At the opposite extreme, there are goats that freshen without enough milk to feed the barn cats.

doe produced 1,300 (590) in 9 months. Notice the lactation curve. The average production goes from 7 pounds (3.2 kg) at kidding to about 8 pounds (3.6 kg) 2 months later. From there it tapers off to about 3 pounds (1.4 kg) at 10 months after kidding.

The lactation curve on page 21 provides another example of an "average" small herd. These are actual, individual records from a herd of four grade does; they show how much production can vary among animals. One doe had a 17-month lactation. She gave 1,800 pounds (815 kg) in the first 10 months and continued to produce a steady 5 pounds (2.25 kg) daily until pregnancy caused production to drop. Another doe reached her peak at 4 months.

If you owned these four does and were going to sell one, which one would it be? There are two lessons here:

First, remember this when you buy a goat: are you buying an animal someone is culling because of low production? Ask to see milk records.

Second, without records and perhaps a chart like this one, no matter how rough, you don't know for sure what's happening. Not a month from now, not a year from now, and certainly not 5 years from now when you're trying to decide which granddaughters of your present milking does to keep and which ones to sell or butcher.

Note that one of these does produced more than twice as much as another, even though they ate about the same amount of feed and required the same amount of care. Note also that two weren't worth milking after only 8 months. And finally, it should be obvious that when you milk your own goats, you don't have as steady a milk supply as when you pick up a

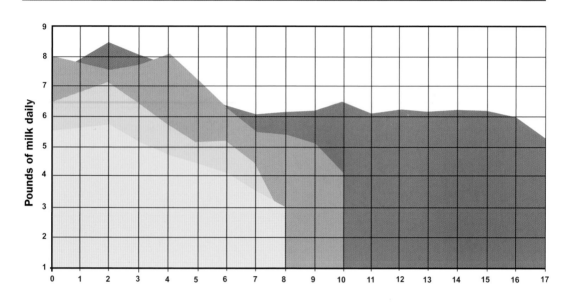

MULTIPLE LACTATION CURVES

Month of lactation

Doe 1
800 lb in 226 days

Doe 2
1,080 lb in 231 days

Doe 3
1,662 lb in 10 months

Doe 4
1,800 lb in 10 months
(17-month lactation)

This lactation curve chart shows how much milk production can vary among goats. The goal of the home dairy is to start out with the best milkers available and then to improve the herd through breeding and selection.

gallon from the grocery store whenever you need it. There will be times when you could drown in milk and other times when you'll eat your cornflakes dry.

Why so much variation? Age is a factor in milk production. Records from this same herd show that peak production comes in the fourth or fifth year. But there are also other factors involved, and any individual goat can vary erratically from one year to the next. Even on a day-to-day basis, milk production is affected by changes in weather, feed, sickness or injury, outside disturbances, and other factors.

The question "How much milk does a goat give?" is best answered by another question: "How long is a piece of string?"

Discovering a New Taste

From all this you can assume that you'll be able to find a goat that will produce a respectable amount of milk for your table for at least a part of the year. Next question: will your family drink it?

Many goat raisers delight in telling stories about how they tricked finicky people into drinking goat milk and how those people couldn't tell the difference. In taste tests, most people will actually prefer goat milk to cow milk. That is not to say that there won't be some people whose taste buds are more sensitive to some of the flavor-bearing proteins in goat milk. They may pick up something unpleasant in the taste and can't get past it. After all, not everyone likes coffee or pistachios. In the same vein, not everyone likes skim milk or whole milk. The texture is different, and the amount of butterfat in the milk will determine how sweet it tastes on the tongue.

There are also those people who taste the milk, find it delicious, discover it's goat milk, and suddenly decide that it's really awful. There isn't much hope for people like this, aside from perhaps serving goat milk from a bottle or carton that came from a cow dairy and not telling them the truth. (I'm not advocating this, but some people have successfully used this technique on their children.)

The best solution is to find a source of fresh goat milk and give it a taste. Remember, I said *fresh*. If you get your first taste of goat milk from a carton of room-temperature, ultra-pasteurized milk — the kind that doesn't have to be refrigerated until it's opened — you are likely to be disappointed.

DID YOU KNOW?

When whipped, cream from goat milk is bulkier than cow cream; the specific gravity of goat milk is 0.83, and that of cow milk is 0.96.

Goat Milk as "Medicine"

When you've been raising goats for a while, you'll surely be asked what you do with the milk. Now we see it in high-end cheeses and gourmet dishes, but it used to be bottled mostly for use in hospitals. The healthful aspect of goat milk is as legendary as goats' aroma and their preference for tin cans but with a more positive spin and a lot more basis in truth.

A few savvy doctors do, in fact, prescribe goat milk for certain conditions. Many more would if pharmaceutical companies didn't make it much easier to dispense a pill or a potion for some of the illnesses traditionally turned over to the healing properties of goat milk. For a time, there was also a problem with getting a steady supply of sanitary goat milk, but that is slowly becoming a thing of the past. Goat milk might be recommended in cases of dyspepsia, peptic ulcer, and pyloric stenosis. It is preferable to cow milk in many cases of liver dysfunction and jaundice because the fat globules are smaller (2 microns versus 2 to 3 microns for cow milk). These smaller fat globules provide a better dispersion of fat in the milk and thus a naturally homogenized mixture (there is more involved in the "natural homogenization" of goat milk than the size of the fat globules, as we shall see shortly).

Goat milk has been used for infants during weaning, people with eczema or other skin conditions, children who are prone to fat intolerance or acidosis, pregnant women troubled by vomiting or dyspepsia, and anxious or elderly people with dyspepsia and insomnia. It is more alkaline than cow's milk and before the days of rampant pharmaceuticals, it was recommended to people with ulcers as a way to help heal the stomach.

LACTOSE INTOLERANCE VS. PROTEIN ALLERGY

The term "lactose intolerance" has become a catch phrase for anybody who gets an upset stomach, gas, or other gastrointestinal discomfort after eating dairy products. Lactose intolerance is the body's inability to manufacture the enzymes that break down lactose, which is a naturally occurring sugar in both cow and goat milk. The fact is, a true lactose intolerance is not common in the general population except in certain ethnic groups. However, milk protein allergies are becoming more and more common. The only way to know for sure is to be tested by a doctor. If the bubbles turn out to be from a lactose intolerance, take it easy on all milk, but if it is a protein allergy, give the goat milk a try.

Goat milk is more easily digested by humans than is cow milk because most of the fat and protein particles are finer and more easily assimilated. People with milk protein allergies — an allergic reaction to the relatively large proteins in cow milk — can usually handle goat milk products. Even those with a lactose intolerance will find goat milk easier on the system. It is particularly rich in antibodies, and when freshly drawn it has a much lower bacterial count than cow milk.

Exploding More Goat Milk Myths

Despite its numerous health benefits, goat milk is not medicine. It's food. Good food! More people in the world consume goat milk, either as a drink or made into cheese or yogurt, than they do cow milk. The United Nations Food and Agriculture Organization estimates that more than 17 million tons (15.4 million metric tonnes) of goat milk is produced — and presumably consumed — annually by just the top 20 developing countries of the world, with India, Sudan, Bangladesh, and Pakistan leading the list. The United States is first on the list of top 20 countries consuming cow milk.

Because most Americans aren't familiar with goat milk, they continue to have misconceptions about it — misconceptions that seem to be based on the comic-strip image of the goat or on the unfortunate experiences of a few who have been exposed to goat milk produced under conditions that make it unfit for human consumption. In fact, many Americans have educated their palates and are seeking out the wonderful flavors of the many kinds of goat cheese now available. But the milk still suffers from bad press. The home winemaker who takes a basket of overripe and spoiled, wormy, moldy fruit, puts it in a dirty crock, and pays no attention to proper fermentation, does not end up with a fine wine. And the goat raiser who milks a sickly, undernourished animal of questionable breeding into a dirty pail, lets the milk "cool" in the shade, and serves it in a filthy glass does not end up with fine milk. While few people would disparage all wines after tasting the concoction just mentioned, many people are all too willing to write off all goat milk after one unfortunate experience.

Milk is as delicate a product as fine wine. It must be handled with knowledge and care, whether it comes from a cow, a goat, or a camel.

What about reports of "odd" or unpleasant tastes in goat milk? One of the more common causes, poor sanitation, is also the most easily remedied. Yes, it can happen, and there are many possible causes. Strict sanitation means more than just "clean enough to eat off." It includes thorough washing of equipment in hot water with a stiff brush, not a washcloth; rinsing with a chemical sanitizer or scalding hot water; avoiding pitted or scratched aluminum containers; milking in a clean, well-ventilated place free of odors; and filtering the milk through an appropriate material (a milk filter made for the purpose, not reused cheesecloth).

Other off-flavors can be eliminated by rapid chilling; keeping milk out of sunlight or fluorescent light; not exposing milk to copper or iron; not mixing warm raw milk with cold or pasteurized milk; and avoiding violent agitation.

Nutritional deficiencies, such as lack of cobalt, vitamin B_{12}, and vitamin E, may also contribute to an off-taste. Some off-flavors can be traced to unsaturated fatty acids in milk that are more susceptible to oxidation, hence causing off-flavors. This can be corrected by decreasing the amount of fat in feed (reducing the amount of soybeans and whole cottonseed, for example) and by increasing the forage-to-concentrate ratio.

Note that in some places in the world, a "goaty" flavor in cheese is highly prized.

In rare cases, animals can give off-flavored milk. We can't call it "goaty" because some cows have the same problem. Off-flavored milk can be caused by strong-flavored feeds and plants such as ragweed, grape leaves, wild onion, elderberry, and honeysuckle. This is one reason many commercial dairies don't let their goats browse or graze: after all, if goats are on pasture, it's impossible to control what the animals eat.

Off-flavors can also be caused by mastitis, which often makes milk taste salty. A rancid flavor can result when a doe is in late lactation or when foamy milk is cooled too slowly (the temperature should drop to under 40°F [4.5°C] in less than an hour). A "cardboard" flavor might be caused by oxidation resulting when milk is left in the light or when there is copper or iron in the milk container. Obviously, a dirty barn, dirty goats, and generally unsanitary practices make milk an excellent environment for bacteria, which can not only affect the taste but also produce illness.

Goat Milk vs. Cow Milk

Goat milk does not taste appreciably different from cow milk. It looks pretty much the same, although somewhat whiter, because it doesn't contain the carotene that gives a yellow tinge to the fat in cow milk. Goats convert all carotenes into vitamin A. Samples of milk with the same butterfat level are equally rich. Goat milk certainly does not smell; if it does, something's wrong.

Most goat raisers enjoy serving products of their home dairy to skeptical friends and neighbors. The reaction is invariably, "Why, it tastes just like cow milk!"

I have noticed, however, that city people who are accustomed to regular standardized milk — milk that has butterfat removed to just barely meet minimum requirements — or, worse yet, who drink skim milk are prone to comment on the "richness" of goat milk. They'd say the same thing about real cow milk if they had the opportunity to taste it before the technologists messed around with it and turned it into chalk water. As with cow milk, the percentage of butterfat — the source of the "richness" — varies with breed, stage of lactation, feed, and age. But generally, there is virtually no difference in taste or richness between whole, fresh cow milk and goat milk.

Biochemical Composition

Despite these similarities, the composition and structure of the fat in cow milk and in goat milk is one of the more significant differences between them.

Although it has long been said that goat milk is "naturally homogenized" because of its small fat globules, it probably isn't the size of the fat globule that causes the cream in goat milk to remain in suspension. Recent research has shown that goat milk lacks a fat-agglutinating protein, a *euglobulin*, that

would cause the fat globules to adhere to one another and mass up. In fact, the cow is probably the only domestic animal that produces milk with this particular protein, according to studies out of the University of Minnesota. Sow and buffalo milk do not form cream lines. Because mechanically homogenized cow milk is the norm in the United States, most people would probably be surprised to see cream rise in milk that comes straight from the bulk tank. They won't be shocked by goat milk, which has very little separation.

Still, from the standpoint of human health, natural homogenization is better. Some research has shown that when fat globules are forcibly broken by mechanical means, an enzyme associated with milk fat (*xanthine oxidase*) is freed. This enzyme can penetrate

AVERAGE COMPOSITION OF GOAT AND COW MILK*		
Element	**Goat**	**Cow**
Water (%)	87.5	87.2
Food energy (kcal)	67.0	66.0
Protein (g)	3.3	3.3
Fats (g)	4.0	3.7
Carbohydrate (g)	4.6	4.7
Calcium (mg)	129.0	117.0
Phosphorus (mg)	106.0	151.0
Iron (mg)	0.05	0.05
Vitamin A (IU)	185.0	138.0
Thiamin (mg)	0.04	0.03
Riboflavin (mg)	0.14	0.17
Niacin (mg)	0.30	0.08
Vitamin B$_{12}$ (mg)	0.07	0.36

*Composition per 100 g

Note: Charts like this can be confusing. There are many similar charts, but seldom do two agree because the composition of milk varies with the animal, the stage of lactation, the feed, and other factors. Averages of one study can differ from averages of another, and neither "average" is likely to apply to your goats.

the intestinal wall in humans, enter the bloodstream, and damage the heart and arteries, creating scar tissue. In response, the body may release cholesterol in an attempt to lay a protective fatty material on the damaged and scarred areas, which can lead to arteriosclerosis. According to Dr. G. F. W. Haenlein, a former Cooperative Extension dairy specialist at the University of Delaware, no such problems are associated with natural (unhomogenized) cow milk or with goat milk.

Goat milk also has a higher amount of shorter-chain fatty acids than does cow milk. Glycerol ethers, too, are much higher in goat milk (which is important for the nutrition of the nursing newborn), and there is less orotic acid, which can be significant in the prevention of fatty liver syndrome.

The protein content of goat and cow milk is fairly similar. Goat milk has more vitamin A and more B_1 and B_3 vitamins (thiamine and niacin) but less B_6 and B_{12}. Cow milk is marginally higher in lactose.

As for minerals, goat milk is higher in calcium, potassium, magnesium, phosphorus, chlorine, and manganese, while cow milk is higher in sodium, iron, sulfur, zinc, and molybdenum. Goat milk has smaller amounts of certain enzymes, including ribonuclease, alkaline phosphatase, lipase, and the xanthine oxidase mentioned earlier.

While there are some minor differences between goat milk and cow milk in nutritional value, in the big picture the differences are inconsequential and hard to find documented anywhere in science.

Raw Milk vs. Pasteurized

If you want to start a dandy (and sometimes heated) discussion in goat circles, just casually bring up the topic of raw milk and stand back.

Most people just buy their jug of milk at the grocery store and don't even think about where it came from or how it was treated. For those of us interested in simple living, however, or even just in producing our own dairy products, it's not nearly so simple.

Here's the problem. Milk is the "ideal" food, for animals, for humans — and for bacteria. Milk is extremely delicate. It can attract, incubate, and pass on all sorts of nasty things such as *E. coli*, *Salmonella*, toxoplasmosis, Q fever, listeriosis, campylobacteriosis, and others that most of us nonmedical people never even hear about. In short, nature's most healthful food can make you very sick.

The Industrial Age answer has been pasteurization — heat-treating the milk to kill or retard most of those threatening organisms. Government regulations now demand that such treatment be performed "for the public good."

This is no doubt a wise policy — for the masses. When you pick up a jug of milk at the store, how would you know if the dairy animals were healthy, if the milker had clean hands or a runny nose, and if proper sanitation measures were taken, without such governmental intervention?

Stainless steel is an ideal container for getting milk from the barn to the house. It is easily cleaned and sanitized and not easily scratched. Pits and scratches make perfect hiding places for dangerous bacteria to grow.

But many goat-milk drinkers raise their glasses with a different perspective. They know everything about their animals, from age and health status to medical history to what they ate since the last milking. The home milker knows exactly how the milk was handled — how clean the milking area and utensils were, how quickly the milk was cooled and to what temperature, and how long it has been stored. (Usually, it hasn't been stored long. Most goat milk from the home dairy is probably consumed before cow milk even leaves the farm, if it's only picked up every 2 to 3 days.)

Under these conditions, many people who milk goats feel it isn't necessary to go to the bother of pasteurizing their milk. Also, many people raise goats because they want raw milk. (It's illegal to sell raw milk, goat or cow, in most states, although goat milk is often sold as "pet food," for orphaned or sickly young animals.) Some think it tastes better. And some say it's more nutritious. There are different considerations when working with cheese and other fermented milk products, but those will be discussed in chapter 16.

Pasteurization does have an effect on nutritive value. But raw-milk opponents claim it's very minor and insignificant when compared to the potential dangers of untreated milk. To them those dangers are horrific: they would just as soon drink poison.

One of the potential problems raw-milk opponents often point to is campylobacteriosis, a gastrointestinal disease caused by campylobacter — a bacteria universally present in birds, including domestic poultry. The symptoms, which range from mild to severe, include abdominal cramps, diarrhea, and fever. Apparently, farm families who drink raw milk regularly build up an immunity. Most of the reported cases have involved farm visitors, those unused to raw milk. Raw-milk advocates point out that the incidence is very small — might as well worry about being struck by a meteorite, they say.

A bacterium much more in the news these days is *E. coli*, which is found in milk or meat that has been contaminated by fecal material, either from dirty human hands or from a sloppy barnyard. *E. coli* is found in the

PASTEURIZATION 101

If you decide to pasteurize your milk, start by pouring the raw milk through a paper milk filter into the container you will use for heating. Slowly raise the milk temperature to 145°F (63°C) and hold it there for 30 minutes or 161°F (74°C) for 15 seconds, being sure to stir or agitate the milk with a sterile utensil so all particles are heated to temperature. A double boiler, if you can find one large enough, will keep your milk from burning. Home-size pasteurizers are available, although they don't usually have a stirring function. (Do not use a microwave, as some people suggest. Dairy scientists have proved that it doesn't work.)

If you prefer raw milk but want an extra measure of safety, have your goats tested for tuberculosis, brucellosis, and campylobacter. And be sure to practice scrupulous sanitation (for more on specific sanitation procedures, see chapter 13).

guts of all mammals and usually doesn't cause a problem in healthy adults, but the highly virulent O157:H7 strain has been implicated in a number of deaths in children that were attributed to contaminated cow milk. Raw-milk opponents don't seem to take the time to differentiate between cows and goats when it comes to condemning the milk.

More serious diseases associated with raw-milk consumption are tuberculosis and undulant, or Malta, fever (called brucellosis in animals). While cattle are susceptible to tuberculosis, goats are highly resistant: they have not been implicated in tuberculosis outbreaks. And while undulant fever is a goat problem in some countries, including Mexico, it hasn't been in the United States.

So who's right: those who oppose raw milk or those who advocate it? Probably both, in certain situations and under certain circumstances.

For example, people with impaired immune systems, such as infants and the elderly, are more at risk for some of the minor diseases that can be passed from goats to humans. In general, my impression is that most people who feel strongly about this, one way or the other, base their decisions more on emotions than on facts. Your personal decision will very likely depend on your psychological makeup: how you regard science and medicine in general, for example, or your attitude toward natural or organic foods, or whether you fasten your seat belt.

Be fair to dinner guests, however, and allow them to make an informed decision about whether to drink raw milk. They might not have the same feeling about their food as you, and as already mentioned, they may not have the same built-up immunities as your family has to minor bacteria hanging around the place.

3

GETTING YOUR GOAT

So you've decided to buy goats. Once you know you have a proper facility and fencing ready for them (see chapters 4 and 5), the next problem is finding them and, perhaps more importantly, finding the ones that are right for you.

The popularity of dairy goats has soared in recent years. The U.S. Department of Agriculture (USDA) estimates that there were more than 375,000 dairy goats in the country at the beginning of 2016, up by 3 percent in just one year. Compare that to the 6.72 million dairy cows in just the top 10 major dairy states, and you'll see that goats are by no means common. And they certainly aren't distributed evenly around the country. There might be many dairy goats in your area, in which case finding the right ones will be fairly easy. Or there might be none at all, which means you'll have to do some traveling to get your home dairy started.

Beginning Your Search

There are many ways to begin your search. If you have friends or neighbors who have goats, or if you have seen goats in your area, you have a good start. Almost everyone with goats has animals for sale sooner or later, and if they don't, they'll probably know of someone who does. They might even know of a goat club in your area, and then you'll have it made.

No doubt you've been watching the classified ads ever since you started thinking about getting goats. Classified ads usually appear for only a few days, so you'll have to read them regularly and probably over a long period. Check out the farm papers serving your area. Again, it might be a lengthy process if there are few goats in your region, but having goats is worth it! Of course, you can also place your own "goats wanted" ad.

THE CASE FOR TWO GOATS

Goats are herd animals. A single goat in the barn will be very unhappy and show it by hollering for attention or acting depressed. Neither shows the true nature of a dairy goat nor endears it to its owner. It's a good idea to start with two goats, even if one is a dry doe or a castrated male. Once you've figured out the routine and are ready to branch out into making cheese, yogurt, and other dairy products, it's time to find another milking doe. A second milk goat can be bred so one is still milking while the other is going through her 60-day dry period. That way, you're never completely out of milk.

Dairy goats are ideal livestock for families. Their relatively small size and generally pleasant demeanor make them popular with children.

Be sure to attend county and state fairs and goat shows. This is often an opportunity not only to see and compare several breeds at once but also to have a good time and to talk to goat people. Reach them at home and they might have a dozen other things to do, but talking about goats is one reason they're at the show! And they might have goats for sale. A show will also give you the opportunity to see what a top-quality dairy goat is supposed to look like. Granted, you are looking for something that will put milk into the pail, not look pretty in a show ring. On the other hand, when it comes to goats, form indicates function.

In most livestock-producing rural areas, you'll also come across "sale barns." Again, the numbers of goats they handle will depend somewhat on how many goats are in the neighborhood, but even if they only see a few every couple of months, they might let you know when they have one or more if you tell them you're interested.

Other local sources to contact for leads on goats include vocational-agricultural teachers, FFA and 4-H leaders, veterinarians, feed stores (if they sell goat feed, the managers will know who buys it), and county Extension agents. Most states have at least one dairy goat association, and many have regional clubs.

Of course, you'll also want to watch the ads in the national goat and livestock publications, such as *Countryside & Small Stock Journal*, *Dairy Goat Journal*, or *United Caprine News*. If you're lucky, you might learn of a knowledgeable and reliable breeder near you. Also, increasingly, the Internet can provide you with information on goats for sale.

These suggestions assume you'll be buying your first goat close to home. When you've gained some experience, you might want a certain breed or bloodline and be willing to buy an animal hundreds or even thousands of miles away, maybe without even seeing it. But at first, it's much wiser and easier to do your searching closer to home. There are several good reasons for that. If you don't know very much about goats, you'll want to see some, and certainly the one you'll be spending a lot of time with. You'll probably want to talk to someone with experience and study her setup, and it's always good to have a go-to person when you think you have a goat health emergency. You'll almost certainly want to make some nearby arrangements for breeding. If you live in a goat-deficient region, shopping close to home might not give you much choice of breed or animals, but that might be more acceptable than traipsing around the countryside (unless you have plenty of time and money).

A pair of Nubian doelings wait their turn in the show ring. Dog-chain collars are appropriate show attire but not safe for everyday wear.

Goat shows and fairs provide good opportunities to see several breeds of goats at once, to learn more about goats by watching the judging, and to talk with goat people.

Terms to Learn

If you're like most people who just want a family milker, you'll end up buying what's available regardless of breed, type, conformation, and even desirable traits. Many people start out like this. It's a good way to gain experience. But you'll still want to have some idea of what to look for, and it will be helpful to at least be familiar with some of the jargon the seller might toss your way. (On the other hand, by the time you finish reading this book, you might know more than the seller. I still chuckle when I recall the lady who showed me her Toggenburgs — which she called "Toboggan Birds." And she was serious!)

You're already familiar with the breeds of goats. But you'll probably encounter other terms that help identify and classify individual goats, such as Dairy Herd Improvement Association (DHIA), registered purebred, Advanced Registry (AR), star milker, grade, recorded grade, American, and linear appraisal. What do all these strange words mean, and how can you use them to help select the right goat?

Dairy Herd Improvement Association

Some dairy goat owners regularly have their animals tested through a Dairy Herd Improvement Association, which keeps records on both goats and dairy cattle. Generally, an independent tester comes to the farm during two consecutive milkings to weigh and take samples of milk from each animal. The samples are analyzed for butterfat, protein, and something called somatic cell count (SCC). The advantage of buying an animal from a farmer on DHIA (or DHI) test is that you know how much the animal has produced without taking the owner's word for it. It also says something about the professionalism of the owner.

A DHIA report is pretty straightforward. It will show the animal's name, when she kidded, how many days she has been in milk during the current lactation, how much milk she gave for that 24-hour test (in pounds), how much she has produced to date since kidding, how much butterfat and protein she has produced (usually in both pounds and percent), and a prediction of how much she will produce if milked for 305 days. Owners have the option of adding other categories and choosing the order in which the categories are listed, but it's fairly easy to interpret the headings at the top of each column.

You don't generally have to ask someone on DHIA to see the records. They are usually proud of the fact that they are on test and will produce them before you even see the goats.

Registered Purebred

A registered purebred animal is one with a pedigree that can be traced, through a registry association's herdbook for the breed, to the very beginnings of the recognition of the breed as "pure."

Some purebreds are not registered, for a variety of reasons. The breeder might not think the goat is good enough to warrant registering. In many cases, smaller breeders who aren't really interested in registered animals simply don't want to bother with the paperwork and expense. In some cases, with a lot of work or luck or both, you can trace a purebred unregistered animal back to its registered ancestors and, with the proper paperwork, have it registered. More commonly, it's impossible to prove that the goat truly is a purebred if it isn't already registered.

If the owner doesn't know because she bought the goat from a friend who bought it from a guy in the next town over, check the goat's ears for tattoos. The American Dairy Goat Association, American Goat Society, and Nigerian Dwarf Goat Association can all trace back recorded tattoos.

Pedigree and Registration Papers

A pedigree is merely a paper showing the ancestry of an individual animal — a family tree. Registration papers are official documents showing that the animal is entered in the herdbook of a registry association.

Advanced Registry

Another term you'll hear is Advanced Registry, or AR. An AR doe is one that has given a certain amount of milk in a year, based on DHIA test. The amount varies with the age of the goat and other factors, but the AR designates the animal as a good milker.

Star Milker

Still another term referring to production is star milker, or ★ milker. It can be earned automatically by having at least three daughters earn their stars or by meeting Advanced Registry requirements. Another way to earn a star is at an official one-day test. The doe earns points for the amount of milk she gives in 24 hours, how many days she has been in milk, and how much butterfat and protein she produces. She must receive a minimum of 18 points to be awarded a star on her pedigree.

Not many goats can earn a star this way; even with lots of points earned for butterfat, protein, and days in lactation, she still needs to give 10 or 11 pounds (4.5 to 5 kg) of milk to meet the minimum. Conversely, many one-day-test ★ milkers can't earn their AR because they don't produce enough in the entire 305-day lactation to qualify. Both are official ratings, however, and both are supervised by someone other than the owner.

If a doe's dam has earned a star and she earns one herself, she becomes a ★ ★ milker. If her granddam also had a star, she is a ★ ★ ★ milker. Bucks can also have stars, signifying those earned by their maternal ancestors.

Grades and Americans

An animal without a pedigree is considered a "grade." It could be purebred, but without the papers you can't prove it. Most grades are mixtures of two or more breeds, and some grades are very good animals. If such a goat meets certain requirements, it might be listed as a recorded grade of whichever breed it most closely resembles.

CALCULATING ★ MILKERS

The star is granted for a minimum of 18 points. Points are based on:

- **Total pounds of milk produced in 24 hours** (one point per pound, measured to one decimal point).

- **Total number of days in the current lactation,** with one point for each 10 days (3.6 maximum for standard breeds and a 1.44 maximum for miniature breeds).

- **Butterfat percentage,** one point for each 0.05 pound (0.02 kg) (multiply the percent of butterfat by the pounds given that day, then divide by 0.05 [0.02]). For example, a LaMancha doe that gives 6.4 pounds (2.9 kg) in the morning and 7.0 pounds (3.2 kg) in the evening would earn 13.4 points. If the doe was fresh for 44 days, she receives 0.4 point for lactation. If the butterfat is 3.7 percent, 0.037 times 13.4 equals 0.4958 pound of fat. This amount is divided by 0.05, yielding 9.92 points (Metric conversion: 0.037 times 6.1 kg equals 0.2257 kg/fat divided by 0.02 yields 11.29. Total points: 17.29). The doe's total points: 23.32. She gets a star!

If such a recorded grade doe is bred to a registered buck, the offspring will be one-half "pure." The does from these matings can be recorded grades. If one of these does is bred to a purebred buck, the kids will be three-fourths purebred. The next generation will result in a goat that is seven-eighths "pure," and one more will produce kids that are fifteen-sixteenths pure.

Seven-eighths does and fifteen-sixteenths bucks can be registered as Americans. You might also encounter Native on Appearance (NOAs) and Experimentals (which usually means a buck jumped the fence and bred a doe of a different breed).

Which Goat Is Best for You?

Assuming you have a choice of goats from all these classifications, which is best? Once again, there is no simple answer. For a home dairy, a registered goat, or even an unregistered purebred, may or may not be the best choice. If you're keen on showing, you'll want registered purebreds. In chapter 9, I'll talk more about the value of looking to purebred animals to improve your herd over time, but at this point our emphasis is on milk.

Registration papers just mean that the goat is listed with one of the registry associations and that the pedigree can be traced back to the closing of the herdbook. This is important to experienced breeders who are trying to upgrade and improve their animals. But if all you want is milk, and you know nothing about the goat's family tree, the papers don't mean much. In fact, some registered purebreds are very poor milkers. A registration certificate is not a license to milk.

On the other hand, many grades, and even brush goats or scrubs, turn out to be excellent milkers. One illustration of this comes from a lady in a Southern state who bought a brush goat. It appeared to be of good Toggenburg breeding, but it had been running semiwild, as brush goats do, clearing a patch of hilly land so cattle could graze there. Brush goats are not fed hay or grain, and they aren't milked. When this particular doe was treated as a dairy goat, she turned out to be a superb milker! She had always had the genetic ability but had fallen into the hands of someone who didn't know or care. With proper feed and management, she blossomed. There are many "common" goats like this.

In other words, you might find very fancy, nice-looking, papered goats that won't produce enough milk to feed the cat, and you might find rather ordinary, crossbred, inexpensive goats that will fill your gallon milk pail to the brim. Your problem as a prospective buyer is, how do you know if a particular goat will be a reliable and efficient milker? This is a two-pronged question. We want to know if a goat will produce milk, and we want to know if she will be efficient, which is to say economical.

Registration papers and pedigrees might tell you something about her milking ability, but only if you know how to read them. Show wins (of the animal herself or of her ancestors) may or may not mean she's a good milker. You might look at stars, AR certificates, classification scores, linear appraisal scores, and more, but a goat that has none of these isn't necessarily a poor milker. It might just mean that the owner hasn't bothered to go after them.

Consider the Price

The upshot is that records and papers give you not a guarantee, but some degree of insurance. And you pay for the insurance. Only you can decide if it's worth it, for any particular animal. If you pay hundreds, even thousands of dollars for a good goat, your milk will be that much more expensive.

But it's even more complex than that. As we'll see in chapter 14, you might lower your milk costs by selling registered purebred kids, if you can sell them for a good price. On the other hand, as a beginner, you might be more comfortable "learning" with a less expensive and perhaps more adaptable animal. To all this add the fact that sometimes there are very good purebreds for sale at very reasonable prices and rather poor grades for sale at exorbitant prices.

Consider the Source

How do you cut through this maze of confusion and conflict? Very much like you buy a car or anything else. You arm yourself with as many facts as possible; you ask some questions; you rely on the integrity and reputation (and probably even the personality) of the seller and the seller's place of business; and then you buy a particular model just because you like the color or because it "appeals" to you!

A BUYER'S CHECKLIST

Here are some questions to ask as you begin to narrow your choices:

- **Do you have faith and confidence in the breeder?** Does the breeder appear to be knowledgeable about goats? Is he trustworthy? You're going to have to rely on that person's experience, honesty, and integrity, to a large degree, for your first purchase.

- **What are the goats' living conditions like?** Are they staked out in a weed patch, or are they well housed, in a neat and comfortable building? Do they have an exercise yard with good fencing?

- **How do the animals look?** Are they disbudded? Neatly trimmed? Do their hooves show obvious care? Are they free of abscesses, patchy ringworm spots, and runny eyes or noses? (Do not under any circumstances feel sorry for the one standing in the corner with its head down and decide to take it home. Chances are it's in the corner because it's ill.)

- **How do the goats respond to the owner?** Do they run away or gather around? Most goats are a little spooked when someone new comes in the barn, but they eventually settle down and get curious. If they stay spooked, you could be looking at a potential problem when it comes time to milk.

- **Are there papers or other official records you can see?** If all the owner has to show you are "barn records," you'll have to rely on the owner's honesty. If there are no records at all, it could mean that the seller doesn't know very much about goats or doesn't care very much about them.

Spotting a "Good" Goat

In all seriousness, it will be helpful for you to have some idea of the differences between a "good" goat and a "poor" one before you purchase an animal. Making this distinction takes a great deal of experience, but following are some general tips to help you get started.

Conformation

The general appearance of livestock, the way an animal is put together, is called conformation. Conformation is what a dairy goat judge is looking at when placing animals in the show ring. While a licensed judge spends many hours and years of study and practice to learn his trade, to a certain degree you must make yourself a "judge" when you buy a goat and as you build and improve your herd.

To become an informed buyer, you must learn the parts of the animal. You must know what good animals look like and what traits are considered defects. You must be observant enough to see both good conformation and defects, and knowledgeable enough to weigh and evaluate their relative importance. Some people have a sixth sense for evaluating livestock, and others never get the hang of it. While you'd have to see and handle hundreds of animals and study far beyond the scope of this book to become even a fair judge of goats, familiarizing yourself with some key concepts will make a big difference when you purchase your first doe.

Shape and Carriage

First, observe the animal from a distance. Note the doe's general shape and carriage. She should be feminine, with a harmonious blending of parts. The show scorecard speaks of "impressive style, attractive carriage, and graceful walk," but this is no beauty contest! These traits are a sign of functionality and can tell a great deal about her general condition, vigor, and dairy strength. That means milk in the pail.

Then move in for a closer look, and don't be afraid to put your hands on the goat.

Head

The head should be moderately long with a concave or straight bridge to the nose, except in the Nubian, which must have a definite Roman nose. Saanens have a concave nose, or dished face. The eyes should be bright, with a broad forehead between the eyes. When you've seen enough goats, you'll realize that the shape of the face is often reflected in the shape of the body. A narrow face often sits in front of a narrow body, and you don't really want either for a good milker.

HOW OLD IS SHE?

Teeth can give a good indication — up to a point — of a goat's age, because they erupt at different times as the goat goes from kid to adult. Pull back her lips and note first that there are teeth only on the lower jaw in front. A baby starts with milk teeth much smaller than her adult teeth. If she has two adult teeth and six milk teeth, she is about a year old. If she has four adult teeth, she is headed for 2 years old, and if she has six permanent teeth she is about 3, and if she has all eight permanent teeth, she is at least 4. From there it gets less precise, but be aware that a goat with heavily worn or missing teeth is probably past its productive prime.

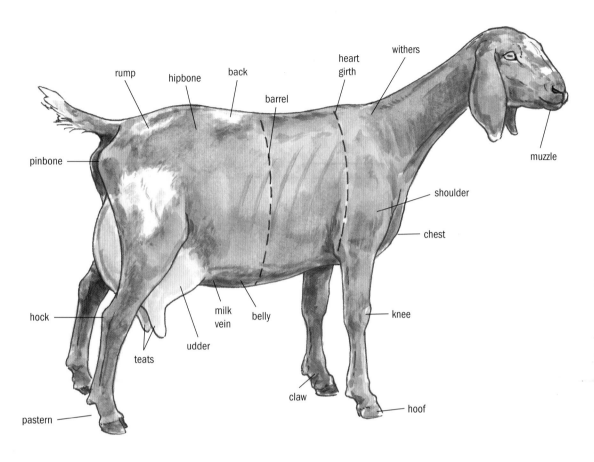

It's important to know the body parts of a goat when discussing the animal with judges, veterinarians, and other breeders.

Ears

Ears are a part of conformation, but they're of small importance if your primary concern is having fresh milk. To make this description more complete, however, note that the ears should be pointing forward and carried above the horizontal, again with the exception of the Nubian, which must have a long, thin-skinned ear, hanging down and lying flat to the head. LaManchas, the "earless" breed, have a size limit of about 1 inch (2.5 cm) per ear. The so-called airplane ears that result from a cross between a Nubian and another breed are ridiculed by many, but again, you don't milk the ears, and more than a few people have such animals and love them.

Muzzle

Of more importance to the home milk supply are such points of conformation as the muzzle, which must be broad with muscular lips and strong jaws, as this is an indication of feeding ability. Large, well-distended nostrils are essential for proper breathing.

Neck

The neck should be clean-cut and feminine in the doe, masculine in the buck, with a length appropriate to the size of the animal. It must blend into the shoulders smoothly and join at the withers with no "ewe neck." The goat needs a large, well-developed windpipe.

Forelegs

The forelegs must be set squarely to support the body and well apart to give room to the chest.

Rib Cage, Chest, and Back

The rib cage should be well sprung out from the spine with wide spacing between each rib. Your hands should feel a flat surface to the ribs, and they should angle back toward the hocks. (A meat goat's ribs will feel like pencils and face more toward the barn floor.) The chest should be broad and deep, indicating a strong respiratory system. The back should not drop behind the shoulders but should be nearly straight, with just a slight rise in front of the hip bones.

Hip Bones

The hip bones should be nearly level with the shoulder. The area between the hip bones and the pinbones should be broad and rectangular but not so long as to make the animal look out of proportion. The slope of the rump should be slight. The broader the rump, the stronger the likelihood that the goat will have a high, well-attached udder — a desirable trait.

Barrel

The barrel should be large in depth, length, and breadth. A large barrel indicates a large, well-developed rumen necessary for top production. If a goat has lots of room to process her food, she has more nutrients available to her system.

Udders and Teats

There are many types of udders and teats. Avoid like the plague abnormalities such as double teats, spur teats, or teats with double orifices.

While very large, so-called sausage teats are undesirable, very small ones may be worse, as they make milking difficult, especially for people with large hands. However, many first fresheners have tiny teats that quickly become more "normal" with milking. It's often easiest to let the kids nurse does like these.

Don't be too impressed by large udders; many of them are just meat. For a pendulous udder, you'll have to milk into a pie pan because there isn't room to get a pail under the goat! Of more serious concern is that pendulous udders are more prone to injury and mastitis infections. If you can encircle the whole top of the udder with your fingers, ask to see a different animal.

A well-attached, capacious udder, carried high out of harm's way, with average-size teats and free from lumps and other deformities, is the heart of your home dairy. Of course, if you can't get the milk out, you've got another problem. Ask the owner if it is okay to take a couple of squirts of milk from each half. You want to see milk without any stringiness that indicates mastitis, and you want to be able to get a full stream without straining your muscles. Ask yourself if this is an animal that you could milk for 3 or 4 minutes without having to call for reinforcements.

Skin

The condition of the skin reflects the general condition of the entire animal. It should be thin and soft and loose over the barrel and around the ribs. A goat with unhealthy-looking skin probably isn't healthy; it might have internal or external parasites. Check for lice, mites, abscesses, or ringworm while you're there.

If you're looking at younger stock, avoid the overdeveloped kid. A kid that develops too early seldom ends up being as good an animal as one that has long clean lines and enough curves to indicate that the framework will be filled out at the appropriate time.

Avoid a skinny goat that is all bones. She won't have the reserve to produce milk. Also, avoid the fat goat that is used to putting her energy into something other than milk production. A goat that shows the outline of ribs and bones and a moderate concavity just forward of the hips is just about right.

Wattles

Many goats have wattles, small appendages of skin usually found on the neck, although they can be just about anywhere on the body. They are a family trait, not a breed characteristic. Some animals of all breeds have them; others don't. They are merely ornaments. Some breeders cut wattles off young kids, not only to make the animal look smoother but because sometimes another kid will suck on the wattles, causing soreness. If wattles are removed with sharp scissors when the kid is just a day or two old, there is little bleeding and rarely even a squawk. You can also remove wattles by tying a thread tightly around the base, which causes them eventually to fall off, but this takes time and can cause some complications best avoided by a quick snip.

Horns

Horns, likewise, are indicative of neither sex nor breed. Some goats have them, some don't. And many goats born with horns have the precursor horn buds removed soon after birth because horns can cause many problems later. Removing horns from an adult is a messy and dangerous procedure. For some goat owners, the natural look of a full set of horns is very appealing, but for reasons you will read later (see chapter 7), it is best to find a hornless goat for your dairy.

Also, these ornaments are a disqualification for dairy goats that are shown.

The Whole Picture

Observe the goat from the side, the front, and the rear, with a critical eye. A good dairy animal has a classic wedge shape when viewed from above and from the side: she has a delicate neck rather than a bull neck, and the barrel is wider than the shoulders. The top line is straight; a severely sloping rump is a defect and can cause problems when kidding. In official judging, general appearance and breed characteristics are allotted 35 points; dairy strength, 20; body capacity, 10; and the mammary system, 35, for a total of 100 possible points. You may not be buying a show animal, but the scorecard indicates function as well as form.

Eventually, you'll come to think of your goat as a living, breathing part of the family, but this is a good time to look at her critically as a machine. Does she have the strong mouth and jaws to take in and chew her food efficiently? Does she have room between her elbows and knees for hardworking heart and lungs? Can her middle carry a full load of food and eventually a kid or two (or three)? Is there room

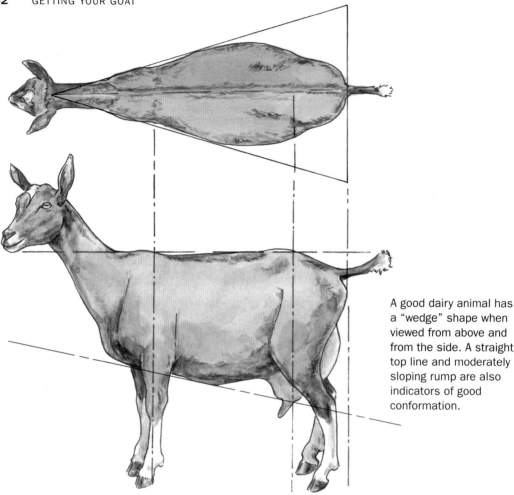

A good dairy animal has a "wedge" shape when viewed from above and from the side. A straight top line and moderately sloping rump are also indicators of good conformation.

between her hips and pins to deliver those kids easily? Is there enough strength and breadth in her rear end to support and walk with a full udder? In a nutshell, does she have the infrastructure to manufacture a product and get it delivered?

Judge the Seller

If for you a "good" goat is simply one that milks well, first consider health, conformation, and overall appearance, because a sickly animal or one that isn't built like a dairy animal isn't going to do the job for you.

Next, consider any records that might be available. The easiest way for the beginner

to judge papers and records, including barn records, is to evaluate the character of the seller. It might be more difficult to judge the owner than the goat, but at least you have a lifetime of experience with people! Without records of any kind, your best assurance of getting the goat that's right for you is by buying from someone you feel you can trust.

Such a person will help you learn, and some will even stand behind the animals they sell (maintaining such a policy sometimes asks a lot, as when careless or ignorant people take home a good goat, neglect or abuse it, and then complain that they were ripped off because the animal doesn't meet their expectations). You want a seller you can call on later for help and advice. When you deal with someone like this, you get much more than just a goat for your money.

On the other hand, there are people who have been raising goats for years and who still don't know as much as you will after reading this book. While some people have 20 years' experience, these folks may have a year's experience 20 times. There are also out-and-out crooks, who will sell worthless animals and overpriced animals. Some will try to sell you "registered" goats with the papers to come later (they never do), and there are others who may be more interested in disposing of a goat or acquiring your cash than they are in helping you or promoting goats.

People who raise goats, in other words, are a cross-section of people in general. If you buy a goat without knowing very much about goats, it will help to know something about people.

There's a reverse side to this, too. Some buyers are pushy, obnoxious know-it-alls. Not you, of course, but bear in mind that an experienced seller has most likely dealt with people like this in the past, and be aware that you are being evaluated, too, and act accordingly.

Official Records and Barn Records

Most people think official milking records such as DHIA are best, but even the owner's own barn records of daily milk production are better than none. Remember that the amount of milk produced in a year is more important than the amount produced on any one day. You don't expect your first goat to break any world records, but you want her to be more than a nonmilking pet, too!

Show wins can be impressive, and they'll tell you something about what qualified dairy goat judges think of the animal's conformation, but like registration papers, blue ribbons are no license to fill a milk pail. Again, registration papers and pedigrees mean little unless you're familiar with a great many names and backgrounds, and that won't come until later, after study and experience. Keep in mind that goat breeders don't usually sell their best animals. Ask the obvious: Why are you selling her?

Be wary of milk records expressed in pints and quarts (and downright skeptical of milk recorded in gallons!). Even an honest and well-meaning milker can be misled by a bucket of foaming milk that "looks" like 3 pints. Weight is much more reliable. Of course, if there is no scale hanging from the ceiling in the milking area, you are right to be suspicious.

Official milking records will be in pounds and tenths of pounds, and so should unofficial barn records. As a quick reminder, a quart (1 L) of milk weighs 2 pounds (0.9 kg); a gallon (3.75 L) is roughly 8 pounds (3.5 kg).

Barn records depend entirely on the accuracy of the scales and the integrity of the milker. They can be falsified or altered. Official tests like DHIA, monitored by outsiders, are much more reliable, but because of cost and other factors, they aren't widely used by goat owners. They are becoming more common in some areas, however. (See page 211 for a typical barn record.)

With or without papers — registration certificates, pedigrees, show wins, barn or official test records, advanced registry certificates or stars — many people recommend that prospective goat buyers see the goat being milked or, better yet, milk her themselves. This is a good time to check for milk flow. With an inexperienced milker, the doe will probably be nervous, and you might not get much milk, but it's better to get a lesson, even if the owner has to finish the job, than to arrive at home and find out that you can't milk. Smell the milk for obvious sour notes that might indicate mastitis or feed problems. I don't generally recommend drinking raw milk from an unknown source and certainly not without knowing that the milk was handled in a sanitary way, but some people may feel it is important to check the flavor of the milk before deciding to purchase the goat.

Assessing a Goat's Worth

You've found your goat, and you're ready to deal. Next question: how much is this goat worth?

BASIC FORMULA FOR FIGURING COSTS

- Add cost of grain times amount eaten per day and cost of hay times amount eaten per day.

- Multiply sum times 365 to get cost of feed per year.

- Divide the cost of feed purchased in a year by the number of pounds of milk produced per year to get the annual feed costs per pound of milk.

- Multiply number of gallons of milk per year (pounds divided by 8 [kg divided by 3.5]) times local retail cost of goat milk to get annual retail value of your milk.

- Subtract feed costs from annual retail value of milk to approximate recoupable price to pay for goat.

Once again, there are no set answers. The price of goats generates as much heated conversation among goat people as anything else. Some plug for higher prices, some for lower prices, and there are good arguments on both sides. Goats have been sold for thousands of dollars, and of course, many more have been given away.

For the person whose primary interest in goats is an economical milk supply, there's a way to determine what an animal is worth. It involves guesstimating how much your home-produced milk will cost. But it will take some basic math, and it isn't foolproof.

Even a formulaic approach isn't really much help. It would be impossible to fill in the blanks of a formula in a book, because hay and grain prices vary widely from one area of the country to another, and so do milk prices. They also vary from year to year; in some places they can double after a drought or crop failure. Still, you need to start somewhere.

An Example

Using the above formula in an example, let's say the local feed mill is selling Purina Goat Chow for $28.40 per 100 pounds, or 28.4¢ a pound. Although we found hay for our beef cattle for $2 per bale, we paid $3.50 per bale for some better hay for the goats; both lots were roughly 50-pound bales. If I feed my goat approximately 3 pounds of grain and roughly 4 pounds of hay per day (with some pasture), over the course of a year I'll use about 1,100 pounds of grain and 1,500 pounds of hay. (Note that hay consumption varies widely, depending on the type and quality; how much pasture, if any, is used; and how much the goats waste. Some people will use as much as 7 to 8 pounds of hay a day, per goat.) The grain will cost me, at current local prices, just a little over $312, and the hay will cost in the neighborhood of $105, for a total feed cost of $417 a year. If this goat meets my expectations and produces 1,500 pounds of milk, my feed costs for the milk will be just under 28¢ a pound ($417 divided by 1,500 pounds), or about $2.25 a gallon (28¢ times 8, the number of pounds in a gallon).

Goat milk in a health food store would cost me (if it were available) more than $11 a gallon, so the 180 gallons or so from my goat would have a retail value of $1,980! Ignoring incidental expenses, and certainly labor, I could pay over $1,500 for the goat and get my money back within a year. The next year that could be considered clear profit, and if she has two kids, I'll do even better.

But wait a minute. Much as I prefer goat milk, what if $11 a gallon is too much to pay, so I opt to match the price of cow milk, which currently sells for $3.80 a gallon? Although we're comparing apples and oranges now, my goat milk would only have a value of $684. I could figure on making $267 above feed costs.

On the other hand, I just heard from one of my nieces who is buying goat milk for her son who had problems with cow milk — and she's paying $4 a quart! Fortunately, he is thriving; unfortunately, I'm in Wisconsin and she's in Utah, so I can't help her out. But at $16 a gallon, she really should get a goat. Plugging that price into the above equation ($16 times 180 gallons per year, minus $417 in feed costs), her "profit" would be a whopping $2,463 a year.

Personal Considerations

There are many other factors to consider. I might not really use that much milk, or my usage requirements might not fit the goat's lactation curve. And of course, never forget that goats are living creatures, not machines. No one can predict if or when they'll get sick, or even die, or how much milk they'll produce.

If economical milk is your real concern, or if you enjoying playing with the spreadsheet program on your home computer, you can spend a lot of time juggling the numbers. By taking into account the life expectancy of the goat and the value of her kids during that period, as well as breeding, veterinary, and other expenses, you can get a fair idea of what a goat will be worth to you. You can also determine how much more you can afford to pay for a goat that produces 500 or 1,000 pounds (225 or 450 kg) more or less than the "standard" 1,500 pounds (675 kg) a year. You'll want to revisit this formula and these considerations when you start to upgrade or when you wonder what effect reducing feed costs would have on your milk bill.

And if you're ever tempted to "go commercial," or wonder why goat milk can cost $14 a gallon in stores, plug in your labor! We won't even go into the cost of commercial equipment. And of course, if goats are your business, you'll have to account for everything from utilities and insurance to legal and accounting services. And don't forget taxes.

In 2008, the Wisconsin Department of Agriculture, Trade and Consumer Protection partnered with Iowa State University Extension to publish a small profitability study of farms that earned the bulk of their income from milking goats and had been in business at least 5 years. It was a small sample, but the "average" farms grossed $32.30 income per hundredweight (cwt) of milk, but it cost them $49.01 to produce it. They actually lost $16.71 per cwt when all costs were factored in! The "top" farms had gross incomes of $31.79 per cwt and spent $28.61 to produce it. They netted a whopping $3.18 per cwt of profit. While the amount of money paid per hundredweight of milk has gone up since the study, so have the costs of doing business.

Be Realistic

The real purpose of bringing all this up at this particular point is to demonstrate that anyone who thinks a decent dairy goat isn't worth more than $25 or $50 isn't being realistic. It costs more — in some places much more — than $100 just to raise a kid to milking age. If you find a seller who doesn't know that, or doesn't care, and will sell you a nice goat for little money, fine. But if a responsible breeder asks for a price in line with what the goat is really worth, don't go into shock. By the same token, if a seller wants to charge what you think is a ridiculous sum, go back over this chapter and assess your situation carefully one more time.

You'll think I'm hedging on giving hard figures, and you'll be right. Time, place, value of the dollar, and demand for milking goats sends the price all over the board. Check the livestock auction prices for goat meat to get the bottom price. Breeders that advertise show animals and big-name breeding stock on the Internet can give you a general idea of what the upper prices are going to be.

In the final analysis, chances are you'll get a goat just because you want a goat and you believe that any goat is better than none. As with a new car, you'll end up bringing home the one that fits not only your budget but also your personal tastes. If your new goat doesn't meet your needs, you might not keep her long, but I can make two guarantees: she will provide a learning experience, and you'll never forget her!

Getting Your Goat Home

Getting the animal home is a matter of using what is available. If you or a friend has a small livestock trailer, that's fine, but I also know of a goat owner whose husband drove semis across the country and would occasionally bed an empty corner with straw to haul home a particularly nice animal. A pickup truck with a topper is ideal. A pickup truck without a topper — even if the animal is cross-tied — is an incredibly bad and dangerous idea.

Goats transport very easily, and many goat owners hauled their first animals home (collared and leashed) in the back seat of the family car. Goats rarely wet or poop unless you travel more than an hour without giving them a break. Just be sure not to dawdle when you stop. Hop them quickly out of the vehicle, and they will leave their deposits in the grass.

Unfortunately, modern times bring us other transport problems that have nothing to do with the animal's comfort. If she is coming from out of state, check with your state's department of agriculture animal health folks to find out what paperwork or health assessments might be needed before bringing her home. There may even be requirements within your state for transporting livestock. While it is pretty easy to sneak a goat almost anywhere, it is not ethical and in some cases could put your whole state's livestock health designation at risk.

4

HOUSING

any people, especially those who haven't had much experience with livestock, are prone to bring home an animal and then decide where and how they're going to keep it. This is definitely putting the cart before the horse. Fortunately, most people contemplating raising goats already have facilities that, with a little work, will serve as a shelter. If you're new to goats, you'd be well advised to learn something about them — not just from books but from practical experience — before building any but the simplest brand-new facilities. A few years' experience will go far toward eliminating costly mistakes.

Most goats today are raised in what is called "loose housing"; instead of being confined to individual stalls or kept with their heads in stanchions like miniature cows, the goats are free to move about in a common pen. While many cow dairies have been converting to this system to save labor, it makes even more sense for goats, because goats are social animals and need companionship. They are also active animals and need exercise. Loose housing is a whale of a lot less work than individual box stalls or tie stalls, and goats can injure or kill themselves if kept on a rope. Loose housing obviously entails lower original cost in both construction time and materials, and it's more flexible. If you have only four stanchions, what do you do when you get a fifth goat?

Ideal Housing

Goats are, in some ways, not awfully particular. They are commonly kept in garages and sheds, old chicken coops, and barns, which may be constructed of wood, concrete, cement block, or stone. The floors may be wood, concrete, dirt, sand, or gravel. Although goat owners like to argue about the relative merits of each, goats do just fine in any of these types of housing.

Ideally, the goat house should be light and airy (but not a drafty wind tunnel) with a southern exposure. It should be convenient to work in, which means the aisles and doorways should be wide enough to get a wheelbarrow through without barking your knuckles; the ceiling should be high enough for a person to safely wield a pitchfork; feed and bedding storage should be conveniently nearby; and running water and electricity should be available to make your work easier, more pleasant, and safer. Seems like most of the features of the "ideal" goat house are more for the benefit of the goat farmer than for the welfare and comfort of the goats!

Housing for a dairy goat need not be expensive or extravagant, but you'll have a healthier, more productive animal if she is able to get out of the elements to a place that is dry and comfortable.

A fairly new type of housing that has some merit for goats is the hoop house, which is usually framed with tubular steel and covered with a heavy, reinforced plastic, sometimes called plastic canvas. This does not refer to the flimsy assembly sometimes used for temporary carports but a substantial building that comes in many sizes and is sold as an agricultural or business structure. Hoop buildings can be purchased with a variety of end coverings or none at all. In many places hoop buildings don't require a building permit, because they are removable if you change your mind.

Hoop housing is airy, is far less expensive than building other types of new structures, and can be converted to shelter something besides goats without a lot of fuss. In fact, hoop buildings make wonderful hay- and equipment-storage buildings. But they have several disadvantages if not set up right for goats. Hoop buildings can act like wind tunnels if not closed off securely on the ends, and the canvas sides must be protected with some

kind of short wall so goats don't chew them or puncture them with their feet. One good way to do that is to sink treated support posts into the ground with 4 to 5 feet remaining above ground. Anchor the bottom edge of the hoop frame to the tops of the posts and run sheets of plywood around the inside of the posts to form a wall. This kind of reinforcement may change the legal status of your structure, so be sure you understand your local ordinances.

In warm climates there is no reason that, with a little imagination, a fairly open hoop building cannot serve all the needs of storage, shelter, and milking area. In colder climates more care must be taken to seal off air gaps and keep piled-up snow from creating a springtime swamp in your animal area.

As for the goats, they do not have to be kept warm even in northern climates if they've been conditioned to the cold through the fall. In any climate, however, their housing must be dry and free from drafts. Goats are very susceptible to pneumonia.

A hoop building supported by a pony wall keeps the canvas safe and shelters the goats with a solid windbreak.

Strand-plastic — thick plastic with reinforcing fibers running through it — can make a good low-cost shelter for goats, but be sure there is adequate ventilation, especially in warm weather when these shelters tend to heat up.

Buildings should be whitewashed or painted white inside. This will make the building more attractive and pleasant to work in, for you and the goats, and light colors tend to discourage flies and other pests.

This chapter includes some suggested floor plans. You'll want to adapt them to your own circumstances. After a few months of doing chores, you'll probably want to make some changes based on your building, your animals, and the way you do things. Again, don't get too fancy or spend too much money right away if you are a novice and don't have any experience with goats.

Flooring and Bedding

Flooring and bedding aren't the same thing, but it doesn't hurt to talk about them at the same time. If you're trying to decide on the ideal flooring for your goat house, it often helps to know what kind of bedding is available, what it costs, and how effective it is for keeping animals dry and warm. Your situation will determine what combination you choose in the end, but the floor and what you put on top of it will contribute a great deal to the health and comfort of your goats.

Bedding Options

The most absorbent bedding is peat moss, which can absorb 1,000 pounds (450 kg) of water for each 100 pounds (45 kg) of dry weight (hundredweight, or cwt), far more than any other bedding material. But if you have to buy it, it's expensive, so few goat owners use it.

LIGHTING THE GOAT HOUSE

Lighting can be used both to increase fall production (16 hours/day starting in September in Wisconsin's latitude, for example) and to induce spring breeding (14 hours/day starting March 1, to simulate the shorter days of fall).

AN IDEAL STARTER SHELTER

This simple, basic, attractive, economical, and practical goat shelter would be ideal for a beginner or anyone with two to three goats for the home milk supply. It uses standard-dimension lumber, has an earthen floor, affords both protection from the elements and good ventilation, and can be built in less than 2 days for $300 to $400 with all new materials.

Its shortcomings include lack of space for storing hay and grain, lack of water and electricity, and no separated space for a milking stand in the shed. These could be overcome simply by expanding the structure or adding to it and installing running water and electricity. But for our purposes this structure works.

Hay is stored on pallets and covered with a tarp. Be sure the tarp is large enough to allow the first bales to be stacked on top of one edge, and then tucked in along the two adjacent sides, with the fourth side firmly weighted down with cement blocks or something similar to prevent wind damage. Besides its minimal cost, this type of storage "shrinks" as the hay is used, so you don't pay for a structure that is half empty half of the time. Just a note about storing hay under tarp: be sure the hay is bone dry before it's covered or it will sweat and become moldy. If you have only newly baled hay available, keep it uncovered when the sun is shining. After a few weeks, it should be dry enough to keep covered.

Grain is kept in a metal garbage can in an adjacent garage, out of reach of the goats.

In good weather the goats are milked outdoors. Milking equipment is washed and stored in the kitchen.

If you decide to keep more goats, this structure could be expanded or used as is for feed storage or kid raising, or it could be converted to a milking parlor. But again, this type of shelter is all you need to get started. And with the experience it will give you, you'll be better able to design your own "ideal" goat facilities. How about a brick, fieldstone, or log goat house? Let your imagination, budget, and experience be your guides.

Chopped oat straw rates second among commonly used materials, but it only absorbs 375 pounds (170 kg) of water per 100 pounds (45 kg) of dry weight. Note that this is chopped straw. Long straw will only absorb 280 pounds (127 kg) per hundredweight. Wheat straw is somewhat less absorbent than oat straw. Wood shavings are highly absorbent, easy to handle, and not as dusty as sawdust. If you have to buy them in compressed cubes, they can be a little too pricey for regular use, but if there is a lumber mill nearby, the owner may be happy to get rid of shavings from the planer. Watch out for shavings from green (unseasoned) lumber, however. Goat kids have been known to get sick and die from eating green shavings with pathogenic bacteria growing in it.

Goats are notorious for wasting hay. They will pick the finest leaves and stems, and let the rest stay in the manger or on the floor. Frankly, if they aren't going to eat it, you might as well pull the old stuff out and use it for bedding (of course, if you are feeding a pig or steer for meat, they will be happy to eat it). Hay isn't particularly absorbent, but it's already paid for and will add another dry layer to the barn.

Cow dairies have put sand bedding to the test and find it is very comfortable for the animals, and it drains well. Goats don't have the same leg and joint problems that their much bigger cousins do on harder ground, but goats also like sand and will search out a sand bank in a field just to roll and rub. Sand can be raked free of "berries" and topped off with dry material, but eventually it will have to be replaced. Then you have the problem of figuring out what to do with a barn full of manure-laden sand.

Bark, wood chips, corn stover (husks and stalks), chopped corn cobs, peanut hulls, oat hulls, dried leaves, and shredded paper have all been used as bedding with varied success. Recently, the proliferation of anaerobic digesters near large dairy farms has made available processed by-product for bedding. Whatever you choose, remember that the goats will try to eat it before they sleep on it, so avoid heavy metals, dyes, or other potentially harmful materials. You will also eventually have to haul it out of the barn, and some bedding products just don't lend themselves to a pitchfork and wheelbarrow. Time and experience will tell you what bedding works best for you.

Wooden Floors

Wooden floors, such as those found in brooder houses or other poultry buildings, can be warm and dry if the rest of the structure is free of leaks. However, wood absorbs urine and will rot. This means that highly absorbent bedding should be used, and it should be changed frequently.

Wooden floors are obviously not the most desirable for animals like goats, where large quantities of wet bedding will accumulate. If you build a new structure for goats, don't put in wooden floors. But if you already have a building with wooden floors, there's no reason not to use it.

CONSTRUCTION TIP

Use screws rather than nails when you build. Screws are stronger, and they come out much more easily when remodeling time comes.

Concrete Floors

According to many experienced goat raisers, concrete floors are only somewhat less desirable than wooden floors. Concrete is cold and sweats when the air is warm and humid. Urine cannot run off, so concrete floors require a thick layer of bedding. I don't consider this a serious drawback unless bedding is very expensive or you don't have a garden to use it on — many people prize the used bedding for use in their fields and gardens or compost bins.

Goat manure and cow manure are quite different. Cow manure is extremely loose and liquid, certainly in comparison with goat droppings. Those neat little compact balls bounce, and in this context bouncing is preferable to splashing! With a little fresh bedding to keep the top surface clean, goat litter can accumulate to a considerable depth and still be much less offensive than that in a cow stall.

The problem with concrete floors is that, while the top of the bedding might be quite clean and dry, the bottom layers can be a swampy morass. Deep litter does not signify a sloppy goat farmer — quite the opposite. For goats, the deeper the better, certainly on concrete floors and especially in winter. The lower layers will actually compost in the barn if they aren't too wet, not only helping to warm the goats' beds in the same way the old-fashioned hotbeds warmed early garden seeds but also hastening its use in the garden. Some people even use compost activator on the bedding to speed up the bacterial action. Such deep litter is warm and quite odorless — until you clean the barn, that is.

Here's a little trick to know if you have an adequate layer of dry bedding. Drop to one knee in the bedding. If you stand up with a wet knee, you need to add more bedding. It's also

a good idea to put your face down at goat level occasionally. If your eyes water or your nose stings, it's not good for the goats either.

Goats prefer less bedding in the warmer months, but even then concrete floors require special management. Concrete is always relatively cool, and hard, and there have been reports of rheumatism-like problems arising among goats on hard surfaces. Sleeping benches — simple raised wooden platforms — can be used both winter and summer, and the goats are enthusiastic about these, regardless of floor type.

Concrete floors have the great advantage of being easy to get really clean, which is particularly important in the summer, when deep litter might not be so desirable. Commercial dairy farmers with large herds often prefer concrete for sanitation purposes. They can be scooped clean quickly with a skid steer, hosed down with a sanitizing solution, and restocked with less likelihood of transmitting disease from one group of goats to another. In the summer, when bare concrete is showing, a hard floor also helps keep goat hooves worn down.

Some years ago there was a lady who kept goats in what was practically the center of town, as the suburbs grew up around her. She kept the animals on concrete floors without bedding. The urine drained away, and the droppings were swept up daily. The place was spotless, there was never a complaint from the neighbors, and the only waste disposal was a daily coffee can of nanny berries that went on the rosebushes.

Dirt and Gravel Floors

I have had concrete floors, and I liked them. But other people have actually torn out concrete to install what they consider the ideal flooring: dirt. That's what I have now. I like that, too, and I have plenty of company.

Earthen floors are the easiest to maintain, as long as you don't dig too deeply when it's time to clean the barn. Excess urine soaks away, and less bedding is needed. This is an important consideration if bedding is expensive and you don't have gardens to use it on. Still, there will be plenty of bedding for the compost pile. Soil is also warmer and more comfortable for the animals than concrete, especially if only a small amount of bedding is used. The one drawback to dirt is that it harbors fly eggs if not cleaned and limed well before rebedding. (Barn lime can be purchased at a farm supply story and is generally sprinkled on a barn floor before rebedding.)

Crushed limestone also makes a good floor, but it tends to pack down over time and become as hard as concrete. At that point it may not drain as well as you want. It can also be very expensive, depending on where you live.

Most goats are kept in a communal loafing area, or loose housing, rather than in individual pens.

Insulation

We'll assume that the roof doesn't leak and that drafts aren't getting into the shelter through cracks in the walls and around windows and doors. We do want ventilation, but not drafts — especially down near the floor, where the goats are.

Insulation often isn't necessary, but it can be highly desirable. It can eliminate condensation, help keep water from freezing, and make the goat barn a more comfortable place for you to work. But you must take special precautions to protect insulation from the goats. Goats will chew, eat, or smash any number of wall materials. Plywood, plasterboard, and the like won't last more than a couple of days. Use stout planks or cement wallboard instead.

Whether or not you insulate, never use a plastic moisture barrier in a goat shed or barn. The plastic won't keep the goats any warmer, and it will cause condensation and humidity problems. A hoop barn may get a layer of condensation inside on particularly cold days, but there is usually enough air circulation to prevent any buildup or dripping.

Size Requirements

The size of the building depends on several factors, including herd size; climate; size and availability of pasture; space for exercise, feed, and bedding storage requirements; and whether you need, want, or can manage a separate milking parlor.

For example, standard recommendations range from 12 to 20 or 25 square feet (1 to 2 or 2.5 sq m) per animal. In warm climates you can go with the lower figure because animals spend more time outside. If there is a sizable pasture or exercise yard and the barn is used mainly as a dormitory, you could also get by with a smaller footprint.

But adequate space for the animals themselves is only one consideration. If you're working with minimum space requirements, don't forget that the birth of kids will require more room in the spring and summer. There is always the possibility that you will keep some of those kids and your herd will increase. Goat herds have a way of doing that, even on well-managed and very small places!

There is also the "Wild West" problem, when one aggressive doe tells another who's lower in the pecking order, "This here barn ain't big enough for the two of us." The worst thing that could happen is to find that one or more of your animals are standing out in the elements because a herd queen is guarding the limited inside space.

AN IDEA FOR DOORS

An overhead garage door works well in a loafing shed. It won't become obstructed by bedding and manure like swinging or sliding doors will, and it can be opened to whatever extent weather conditions dictate. Used garage doors are often available for little or nothing. Just be sure they can be set with some mechanical stop to keep the goats from sliding them open and closed on a whim. A sideways sliding door needs the same kind of antigoat brake. To keep drafts out and bedding in, screw a 2 × 10 plank across the bottom of a sliding door opening in winter, and open the door just wide enough for one doe to go through at a time.

You'll need space to store hay, bedding, and grain, and more likely than not the milking will be done in the barn, too. Technically, it should be in a separate section, away from dust, manure, and odors, but most people with only a few goats find this hard to justify. For them, milking outdoors in nice weather can be an acceptable substitute, but milking in a barn aisle is common.

In other words, if you grow your own hay and straw or buy it off the baler when it costs less and plan to store a year's supply, you'll naturally need more room than if you intend to bring home a week's supply at a time. If you spend more money to have a larger building, you can hope to save some by buying feed and bedding in large quantities. Note, though, that if you stock up on hay and straw during the harvest, roughly half of this will be gone by kidding season, so consider planning a combination hay-storage and kid-raising area.

Here again, if you already have facilities that aren't perfect, don't panic. If there simply isn't room for hay in the building you'd like to use for goats, it can be stored elsewhere. It will involve more labor, but it won't keep you from raising goats. Grain can be stored in garbage cans in outbuildings, if necessary. Milking elsewhere can be extremely inconvenient, however, especially in inclement weather. Unless an alternate shelter is very close to the goat shed, try to incorporate milking space into the main building.

Grade A dairies, cow or goat, must have a separate room for handling milk, and a separate milking parlor is advisable to eliminate dust and barn odors. But for the backyarder or homesteader with a few animals that are kept clean, milking in an aisle is acceptable, and far more convenient. Also, a milking bench isn't a necessity, but it will contribute greatly to the ease of milking and will result in a superior product.

The Manger

Never just throw hay or grain on the ground — use a manger to feed your goats. Grain is often fed in mangers, since most goats won't get their allotted ration during milking. Greed can cause problems, unless some means of fastening the animals into the manger is devised to prevent bossy does from taking more than their share. Hay can be fed free choice in the same manger; the goats can come and go as they please and eat as much as they want.

There are many styles of mangers, but the best ones consider the nature of the goat. Goats are notorious wasters of hay, and this is the main factor to consider when designing and building a manger. It should also be difficult for goats to climb into or otherwise contaminate, and of course it should be easy for the goats to eat out of and for you to clean and fill.

The keyhole manger, a longtime favorite, has the shape of the classic keyhole, with a 7-inch (18 cm)-diameter hole at the top and a 4-inch (10 cm)-wide slot coming down from that. The doe must reach up to get her head through the hole, and then slide her neck down the slot. If you want to constrain a goat in this device, perhaps to trim hooves or even for milking, a simple latch to keep her neck in the slot will do it.

Based on his half a century of experience with goats — as many as 1,000 at a time — the late Harvey Considine devised a modified version of the keyhole manger. It has two unusual features that deal with goats' habits.

First, the goats eat while standing with their front feet on a "step" a few inches above the floor level. And to get at the feed they must put their heads through openings made of slats or bars set at an angle of 63 degrees. The openings range from about 4 inches (10 cm) for small kids to 5 inches (13 cm) for mature does to 7 inches (18 cm) for large bucks. The animals must then reach down to get the feed.

The angle is important. The goats can see the hay and can get at it by tipping their heads. Ordinarily, goats would grab a mouthful of hay and back off, dropping half of it on the floor and refusing to touch it again. But the angled bars do away with this problem. If they back out quickly without turning their heads, they will give themselves a nasty clip on the ear. Within a few days you will find them standing in a line with their heads in the manger until they are done eating. No more grab-a-bite-and-run problems. It's a beautiful sight, especially if you're paying several dollars a bale for hay!

You can use 1 × 4 boards for the slats, but these obstruct the goats' view of the feed and aren't very durable. Instead, Harvey recommends steel strapping, $\frac{1}{8}$ inch (3.2mm) thick, 1½ to 2 inches (3.8 to 5.1 cm) wide, and about 24 inches (.61 m) long. The bottom feed trough can be any width, but a goat won't be able to reach more than about 16 inches (40 cm). These should be about 42 inches (1 m) high on the goats' side to keep them from jumping into the manger.

The manger is best built as a fence or divider inside the barn so you can fill and clean it from the aisle. It can be built along a wall, of course, but that will make your chores more difficult.

Goats are notorious for wasting hay. A well-designed manger can minimize this problem.

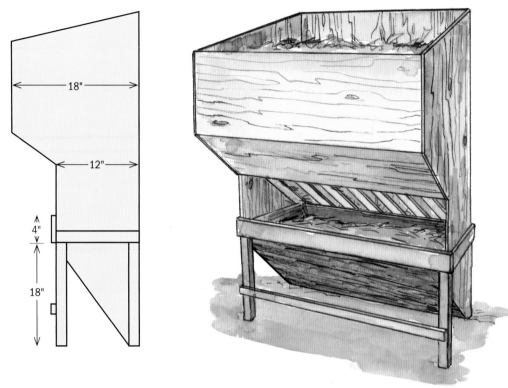

This design for a hay manger is simple yet functional. Make it as long as necessary for the number of goats you're feeding, allowing 18 inches (45 cm) or more per animal.

ANOTHER FEEDER IDEA

Sections of polyvinyl chloride (PVC) pipe, cut in half lengthwise, make good feeders for grain or loose minerals, or even milk. A 3- to 4-inch (7.5 to 10 cm)-diameter pipe works well for kids; mature animals need a pipe about 15 inches (40 cm) in diameter. Make the feeders as long as you need. Block the ends with PVC caps, and hang the feeders with metal strapping. The PVC is easy to clean. Remember, anything that hangs inside the pen becomes playground equipment for goats to climb on, which means contaminated feed. Whenever possible, keep water supply, feed pans, and mineral supplies outside the pen with openings that the goats have to reach through with just their heads.

Gates, Latches, and Fences

Gates and latches are important in goat houses. For the goats, they serve the purpose of entertainment. For the owner, they are supposed to keep goats in or out of designated areas. Gates should be sturdy, for goats love to stand on things with their front feet, and gates are the favored place to do this standing. With deep litter, the gate should swing out of the pen, which is a good idea in most cases anyway. Make sure the gate is wide enough to get through with a wheelbarrow or whatever you'll be using at cleaning time. Sliding gates are very nice, but more difficult to build, and they get stuck as bedding builds up against them.

As to the ideal height for fences: goats vary widely in jumping ability, or perhaps desire. Contented goats are less likely to leap fences of any height. But if a deep litter system is used, remember that a 4-foot fence in October might only be 3 feet high a short time later! (And naturally, as the floor goes up, the ceiling comes down, an important consideration if it's already low or if you're tall.)

Since most goats are Houdinis when it comes to unlocking latches, pay special attention to those. A double-jointed eye hook is a good choice.

Other Considerations

Some people demand more luxuries and convenience than others, whether they are goat raisers or not. For goat folks, running water and electricity in the barn are often the first items on a wish list. Extra storage space is usually next.

Utilities

Water piped to the barn can save countless minutes, which on an annual basis amounts to hours or even days. The goats are more likely to have a continuous supply of fresh water if you don't have to lug it a long distance, and from that standpoint alone the plumbing can be worthwhile. A hose might work in summer or in a warm climate, although it's an unsightly nuisance and the water can get quite hot. Where freezing occurs, buried pipe and a frostproof hydrant can almost be considered necessities.

Sturdy gates and secure latches are of special importance in goat barns. Goats are amazingly adept at opening ordinary hooks and latches.

CAUTION!

Be certain that wiring and cords are not exposed for the goats to chew. *Anything* that sticks out, hangs down, or offers a gap to get stuck in will attract a goat like bees to honey. Even if it doesn't look like they can reach it, at least one of them will. Never leave loops of baling twine, extension cord, hose, or lead rope hanging anywhere near a goat pen. There will be at least one goat that will try to figure out how to get it around its neck.

Electricity is obviously a boon when you must do chores before or after the sun shines. Trying to milk or deliver kids by flashlight is challenging, to say the least, and lanterns can be dangerous as well as a bother. Moreover, eventually you'll want electricity for clippers, disbudding irons, heat lamps, and perhaps for a stock-tank heater to keep drinking water from freezing, as well as other possibilities. If you need more reasons to run electricity to the barn, some people say a radio tuned to a classical music station makes their goats give more milk.

Storage Space

You'll want storage space, of course. How much you need depends on the type of operation you have. There should at least be room for a pitchfork that can be kept out of harm's way, hair clippers, hoof-trimming tools, brushes, disbudding iron or caustic, and a medicine cabinet. Provide a place for a hanging scale, and make sure to keep the milk records where the goats can't nibble on them (this is the voice of experience speaking)!

Milking equipment must be stored somewhere cleaner than the barn, of course. The ideal milk house — well ventilated, with hot and cold running water, rinse sinks, floor drain, and impervious walls and ceiling — is nice, but it's a bit much to expect for a dairy with only a few goats. The kitchen works just fine for most people. That's where the utensils and strainer pads are kept, the milk strained and cooled, and all milking utensils washed.

Final Thoughts

Facilities for dairy goats need not be elaborate or expensive. But because goats are dairy animals, you'll want to keep them and their surroundings as clean as possible. Plan quarters that are easy to keep clean, that are pleasant for both you and the goats to be in, that have been "kid proofed" from curious goats, and that will contribute to the health and well-being of your herd. In the end, the design of your barn depends on your building site, budget, individual situation, and personal preferences. There is no "best" plan for everyone. If a design meets your needs, then it's a good one.

A TIP FOR KEEPING FACILITIES CLEAN

At least once a year, scrub down mangers, walls, and concrete floors with a bucket of hot water containing 1 cup each of bleach and cider vinegar. Dirt floors should be dusted liberally with barn lime, paying special attention to corners and anyplace that tends to retain moisture. Lime helps control odor and pathogens associated with manure.

SAMPLE FLOOR PLANS

18 × 18 Goat Barn

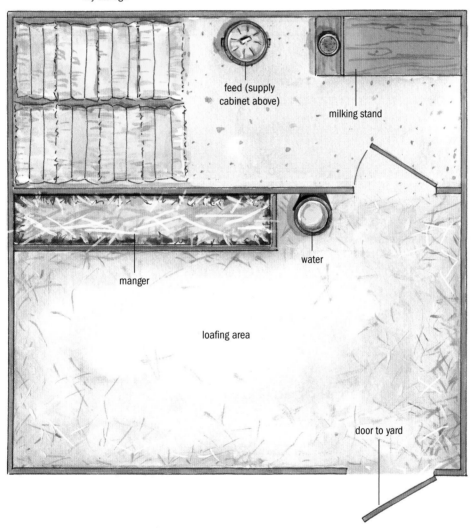

hay storage

feed (supply cabinet above)

milking stand

manger

water

loafing area

door to yard

This is a good basic floor plan, showing the fundamentals of a goat barn. Grain is stored in a metal garbage can with a tight lid. Many alternative arrangements are possible, including locating the water outside the loafing area to help reduce contamination and providing a wider door to the yard to prevent bossy does from blocking the entrance.

SAMPLE FLOOR PLANS

8 × 15 Shed

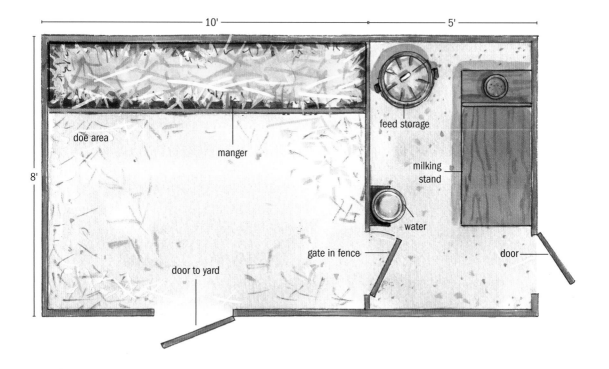

Although a very small shed is workable, it can hamper efficiency. This plan would require carrying hay from an outside storage area all the way through the shed to reach the manger. Sliding doors and a folding milking stand (see page 194) are especially valuable in cramped quarters like this.

HELPFUL HINT

Hang a toilet scrub brush near the water faucet, and clean water buckets every time you fill them.

Barn

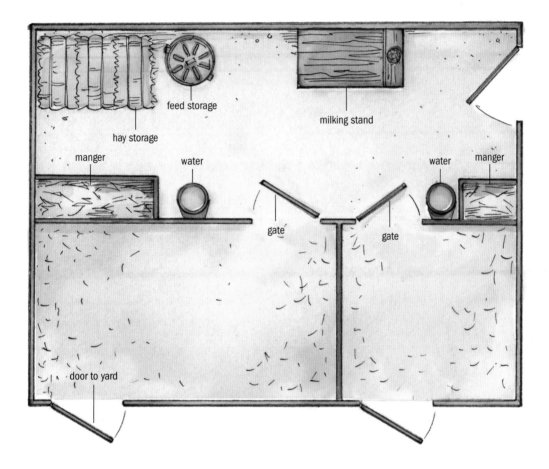

Barns should be designed for efficiency. They should be easy to clean and keep clean. The layout should eliminate unnecessary motion; function is more important than appearance. For instance, in this design one pen can be used to hold goats that haven't been milked. As each one is finished, she can be shuttled into the empty pen.

If you only have a few goats and one shed and want to keep a couple of kids, this arrangement could work just fine. The smaller area might also be used as a kidding pen or for hay storage.

5

FENCING

Fencing is probably more important — and more difficult — with goats than with any other domestic animal. One professional fence builder said his way of testing a fence for goats is to throw a bucket of water at it. If the water can go through, so can a goat. Goats will jump over, crawl under, squeeze through, stand on, lean against, and circumvent any boundary that is not strictly goatproof. Even if it is goatproof, they'll spend time trying to figure out how to make it fail. Fencing is important not only to keep goats in, but also to keep stray dogs and other predators out.

How Much Is Necessary?

With goats, a little fencing goes a long way. In most cases, if you have only a couple of goats, you won't want to think in terms of "pasturing" to any great extent. Goats won't make good use of the usual pasture plants, grasses, and clovers. They prefer browse: trees, shrubs, and brush. Goats that are fed at the barn will probably ignore even the finest pasture, although they'd be delighted to get at your prize roses, specimen evergreens, and fruit trees. For many people, protecting valuable plants like these is the main reason for good fences! Goats also like to jump on cars and other machinery, so make sure vehicles and goats are kept apart.

We'll talk more about pastures and pasture fencing later. For now, let's focus on the exercise yard. A small, dry, sunny yard adjacent to the barn is all you need, ordinarily, and you'll probably want one of these even if you pasture your animals. The exercise yard fence will take more punishment than the average pasture fence, because the goat confined to the smaller space will have more time and opportunity to investigate and beat on it. The cost per running foot will be higher in the yard, but the amount of fencing used is much less.

What Kind of Fence?

Good fencing is obviously a necessity for goats. And while fences do require an investment, the newer types make it much easier and more economical to allow goats access to larger yards or pastures. First, we'll discuss the fences to avoid, followed by the ideal types and some practical alternatives.

Stock panels are made of ¼-inch (0.64 cm) welded steel rods, which make them sturdy and ideal for goat pens. The standard length is 16 feet (about 4.75 m); the 52-inch (1.3 m)-high ones with narrower bottom openings work well for goats. A recent improvement features panels with 4-inch (10 cm) spaces, which eliminates the problem of horned goats getting their heads stuck in the fence.

Types of Fencing to Avoid

For various reasons, these fences are not recommended.

Woven wire. This type, also known as field fencing, is less expensive but has drawbacks. If your goats have horns, they'll put their heads through the fence and then be unable to get free. Worse, they'll stand on the wire, or lean against it, until it sags to the ground and they can nonchalantly walk over it. Even with close spacing of posts and proper stretching — a crucial part of building this type of fence — woven wire will soon sag from the weight of goats standing against it and will look unsightly and eventually be useless. However, woven-wire fencing can be useful when combined with electrical fencing (see page 71).

Rail fence. For many people, the picturesque board or rail fence comes to mind first, but it won't work for goats unless it's all but solid. They can slip through openings you wouldn't believe. Don't take the chance.

Woven-wire field fencing is relatively inexpensive, but goats can easily ruin it by standing on and leaning against it. It can be constructed with wood or metal posts.

Barbed wire. It's awfully ugly stuff around tender-skinned, well-uddered goats. And it doesn't impress them anyway. Some people use barbed wire, often because they feel it deters predators. Far more would suggest getting rid of it.

Picket style. Another nasty trap for goats. With a picket-style fence, there's a very real danger that a goat will stand against the fence with her front feet, slip, and impale her neck on a picket or get it caught between two.

Useful Types of Fencing

Here are better, safer options that work.

Chain-link. This is the ideal goat fence for a small place. However, like most ideals, it may simply not fit the budget.

Stock fencing. A very good and somewhat less expensive alternative to chain-link is stock fencing, often called hog or stock panels. It is made of welded steel rods in 16-foot (4.75 m)-long panels of 3-foot (almost 1 m) or 5-foot (1.5 m) heights. The ideal is 13-wire combination fence sections that are 5 feet high and have narrower spacing at the bottom than at the top. They do a better job than the standard livestock panel at keeping small kids inside. Attach stock fencing to regular steel or wooden fence posts. They can be connected to each other with small cable clamps, or make temporary pens by connecting them together with electrical zip ties that can be cut loose quickly when you are finished with them.

Sheep stock fencing. Regular stock fencing has one problem: horned goats can get their heads trapped between the rods of the wider openings. A newer and better version has smaller spaces (about 4 inches [10 cm]) between the rods. This makes it more expensive, but it's much safer for your goats. This type of fencing was designed for sheep, and unfortunately, it isn't available everywhere.

Electric fencing. This should be used much more than it usually is for goats. They must be trained to respect it, but once they know what happens when they touch it, it's possible to fence even large areas at low cost. Train the goats in a small area. Until they get zapped once or twice, they'll be trying to crawl under, jump over, and just plain bust right through. Unlike other livestock, a goat will jump forward when it gets a shock.

Chain-link
fencing is ideal
for goats, but
it's expensive.

Stand by during the training process to toss the goat back in the pen for another training session. With a good hard shock supplied by a low-impedance New Zealand–style charger, or fencer (see below), it usually takes only one reminder for a goat to learn which side of the fence is home. Do not assume they will quit looking for ways to escape, though. They have an uncanny sixth sense about electricity and will test the fence again when they think the power is out. Most likely, the older does will send a kid to test the fence while they stand back and watch. You might want to use sturdier (and more expensive) fences for smaller yards, but electric fences are ideal for larger areas such as pastures.

You can also use electric fencing in tandem with woven-wire, or field, fencing. Place a strand of electrified wire just inside the woven wire, at about goat-nose height. This combination makes a very good goat barrier: the electric fence keeps the goats from reaching the field fencing, and the field fencing offers more security than the hot wire alone.

New Zealand–style. A fencing system that has become widely popular was developed in New Zealand for managed intensive grazing. Fences can be charged with a 12-volt battery and energizer, a plug-in energizer, or a solar-powered energizer. The most important thing is that the system is very high voltage — 5,000 to 7,000 volts — for a fraction of a second. It won't do serious damage to the goat or predator, but it leaves a lasting impression. The fence itself is usually high-tensile wire supported by wood corner posts and fiberglass spacers. The charger is a little pricey, but the fence is less expensive per foot than any other option suitable for goats. One option for creating a temporary fence is plastic step-in fence posts, a reel of polywire (a thin plastic cord braided through with seven or nine strands of wire), and a solar charger. Some companies also make an electrified woven netting for temporary fences. Goats can be turned into an area to browse without your having to build a permanent fence. **Be sure to train your goats to this type of fence, and watch carefully for the first few days to be sure nobody gets snagged in the soft netting.**

An electric fence carries a pulsating (not steady) current provided by a fence charger, which can operate on household current or a battery. Solar-powered models are available.

A SAMPLING OF TYPES OF FENCING

There are two types of fences: a low-voltage zapper that is less expensive but easily shorted out by weeds, and a low-impedance charger that can cost a couple of hundred dollars but has enough voltage to carry through most weeds and fallen branches over long distances with a good sting.

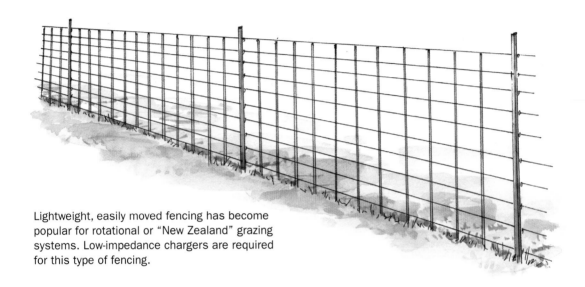

Lightweight, easily moved fencing has become popular for rotational or "New Zealand" grazing systems. Low-impedance chargers are required for this type of fencing.

Adding an electrified wire will keep goats from standing on the fence. This is more important with flimsy woven-wire or field fencing than with the sturdy stock panels shown here.

FEEDING

No aspect of goat raising is more important than feeding. You can start out with the very finest stock, housed in the most modern and sanitary building, but without proper feeding your animals will be worthless. And feed is the goat owner's biggest expense.

The proper feeding of goats requires special emphasis for several reasons. High on this list is the fact that many people who start to raise goats have little or no experience with farm animals. Feeding goats is a lot different from feeding cats or dogs or parakeets. Goats are ruminants, which affects their dietary needs. And unlike the average cat or dog, goats are productive animals, which puts an additional strain on their bodies and requires additional nutriment. To top it all off, they are very picky eaters and won't necessarily eat what you spend time and money getting into their feeder.

The bulk of a goat's diet consists of forages: green plants or hay. But this doesn't mean that just any old grass will provide the nutrition a healthy, milk-producing or pregnant goat requires. So we'll take a look at some of the more common pasture plants and what every goat owner should know about hay.

Browse and hay do not provide all the vitamins and minerals the productive animal needs either. The concentrate, or grain ration, provides these. We must know at least the basics of this, too.

But before we talk about specific feeds, let's weigh some of the considerations involved in selecting a feeding program.

Proper feeding of goats — for their well-being and for milk production — requires more than turning them out into just any old grassy area. This chapter explains why and how to do it right.

Goats will certainly graze with their heads down if that is where the best forage happens to be, but they are browsing animals by nature. They prefer to have their food delivered in a manger that is at chest height or above.

Hay is mixed grass or alfalfa that is cut in the field, pressed and tied into bales, and stored for later use.

The Long and the Short of It

A discussion of feeds can be very long or very short. It can be short if you buy good hay and a commercially prepared grain ration and follow the directions on the label. That's what most goat owners, especially beginners, will do. The discussion on feed will have to be long if you grow hay and mix your own concentrates, because that will require at least an awareness of the elements of nutrition, physiology, bacteriology, math, and more. It will also be a little longer if you want to understand what's involved in buying those off-the-shelf feeds. Understanding that will make you a better goat raiser and give you a greater appreciation of your goats' nutritional needs.

We'll discuss pasture, hay, and grain. But to understand their importance, first we should know something about the animal's digestive system.

The Digestive System

People familiar only with human diets and perhaps those of dogs and cats should especially examine the process of rumination, because goats are ruminants. Like cows and sheep, goats have four "stomachs."

The process of rumination serves a very definite purpose and has an important bearing on the dietary needs of the animal. Ruminants feed only on plant matter that consists largely of cellulose and other carbohydrates and water, making adaptations in the structure and functioning of the stomach and intestines necessary. We commonly speak of "four stomachs," but in reality, the large rumen (or paunch), the hay reticulum, and the omasum ("many plies") are all believed to be specialized segments of the esophagus, while the fourth stomach, the abomasum, is the true stomach and corresponds to the single stomach of other mammals.

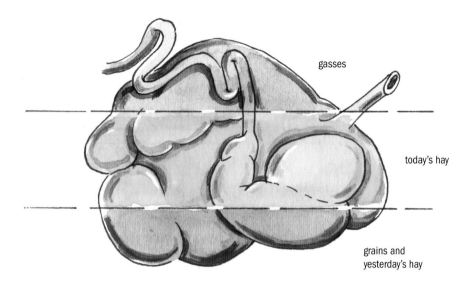

gasses

today's hay

grains and
yesterday's hay

The top of the rumen is filled with gas, and the middle contains
recently eaten hay, which floats atop the bottom slurry of
yesterday's hay, grain, and fluid.

Vast numbers of protozoans and bacteria live in the rumen and reticulum. When food enters these "stomachs," the microbes begin to digest and ferment it, breaking down not only protein, starch, and fats but cellulose as well. The larger, coarser material is periodically regurgitated as the cud, rechewed, and swallowed again. Eventually the products of the microbial action (and some of the microbes themselves) move into the "true" stomach where final digestion and absorption take place.

No mammal, including the goat, produces cellulose-digesting enzymes of its own. Goats rely on the tiny organisms in their digestive tracts, introduced by their environment, to break down the cellulose in their herbivorous diet. The microscopic organisms and the goats have a symbiotic relationship: the organisms help goats digest their feed into usable nutrients, and the goats provide the organisms with food. You might say you're feeding the microbes rather than the goats, because without the microbes the grass and hay would have no food value.

These tiny organisms, conditioned as they are to a given diet, can't cope well if their diet is altered. The result of a diet change is usually a sick goat. Therefore, make any feed changes gradually. Many a goat raiser has fed a goat an

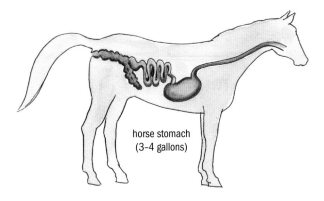

horse stomach
(3–4 gallons)

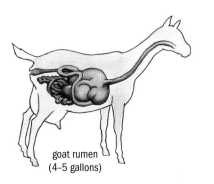

goat rumen
(4–5 gallons)

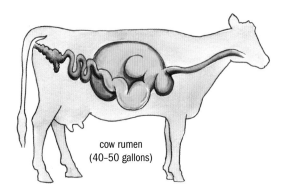

cow rumen
(40–50 gallons)

As an illustration of why the feeding requirements of ruminants differ from single-stomach animals, note that a horse's single stomach holds 3 to 4 gallons (11.5 to 15 L), while the rumen of a well-developed but much smaller goat has a capacity of 4 to 5 gallons (15 to 19 L). A cow's rumen can hold as much as 40 to 50 gallons (150 to 190 L)!

armload of cornstalks salvaged from the garden after harvesting sweet corn, and when the goat gets sick or dies, the cornstalks get the blame. In fact, the problem was overload. Feed such delicious things sparingly, along with the regular diet, and everybody — protozoans, goat, and you — will be much happier.

Let's back up a bit to take another look at these stomachs, for not only are they of obvious importance to the goat, but the goat owner (or at least the kid raiser) has some control over their development.

Developing the Rumen

The rumen and reticulum together occupy about 30 percent of the stomach space of a young, milk-fed kid. At maturity, a well-developed doe has a rumen that occupies 80 percent of the stomach space and a reticulum that takes up 5 percent. The goat must have a well-developed rumen to function properly, and she requires a bulky diet to keep the rumen working properly. The rumen does not increase in size without proper stretching or development; therefore, early feeding of roughage is essential and remains the basis of the goat's diet.

If you watch a mature doe eat, you'll see that she takes little time to chew. She draws in her neck to swallow, allowing the food to slip through the slit in the esophagus to the rumen. A slight fermentation begins as the microbes go to work. When at leisure, the doe regurgitates some of this material and "chews her cud." To swallow the now thoroughly masticated food, she extends her neck, and the cud goes to the third stomach, or omasum. When she drinks from a pail of water, she extends her neck to the far side, ensuring that the fluid goes to the omasum where it belongs, not to the rumen.

HOW A KID DRINKS

Watch a newborn or very young kid sucking. She stretches her neck out to get her milk. Because of this stretching process, the milk can move past a slit in the esophagus, bypassing the first two stomachs and ending up in the omasum. Here it is mixed with digestive fluids and is passed on to the fourth stomach, or abomasum.

Contrast this with a pan-fed kid, especially one fed only two or three times a day instead of four or five and who is therefore more hungry and greedy. She must, first of all, bend down to drink rather than stretch upward. Some of the milk slops through the slit in the food tube and falls into the first stomach, the rumen, where it doesn't belong. There is nothing else in this compartment, since milk is the only feed consumed. There is no bulk. Gas forms, and scours is likely to result.

The good goat raiser strives to keep milk out of the rumen by proper feeding. Moreover, the breeder will work to develop the rumen and reticulum the way they should be developed by encouraging the kid to eat roughage at an early age.

Feeding for Milk Production

In addition to the special needs of the goat relative to rumination, it's important to feed her as a dairy animal. Production of milk requires more protein than would be needed just for body maintenance, for example. So a milking doe is fed a ration of at least 16 percent protein, while a dry mature doe or buck will do well on 12 percent. Protein is expensive, and any excess is just wasted. You want to make sure the diet has enough but not too much. Dairy animals have a greater need for calcium and certain trace minerals as well.

BULK AND THE BRITISH POSTWAR EXPERIENCE

Excessive feeding of milk and concentrates to young goats apparently prevents full development of the rumen; advocates of early weaning agree.

In his classic book, *Goat Husbandry* (London: Faber and Faber, 1957; 5th edition, 1993), David Mackenzie maintains that bulk is necessary for good milk production. He points out that milk production in British goats dropped by 12 percent just in the 4 years following postwar "derationing" of animal feedstuffs in 1949.

When concentrates were rationed during the war years, the official concentrate ration for a milking goat was adequate if she had plenty of bulk food such as hay and roots. The allowance for kids and young stock was much more restrictive, and milk for kids even more so. Despite this, Mackenzie's charts show a steady increase in milk production during rationing, based on about 3,000 records from the British Goat Society, and a dramatic decrease in milk production after rationing was lifted.

Basic Nutritional Requirements

It might be very helpful to think in terms of minimum daily requirements, which most of us are familiar with nowadays for humans. Goats, too, have minimum daily requirements. Remember this, and you'll be less tempted to stake the animal in a brush patch and assume she's "fed" just because she filled her belly. She has no more nourishment in that situation than you would if you lived on candy bars and soda pop.

Most of the early information on goat nutritional needs came from extrapolations of data from cattle studies, but a goat raiser will be quick to point out that goats are not little cows. As goats have moved into the realm of commercially important livestock in some parts of the country, research has followed, especially at the E (Kika) de la Garza Institute for Goat Research at Langston University in Oklahoma. There is more information on the nutritional requirements of meat goats, but dairy goat nutrition is quickly catching up.

Undoubtedly, what we know today will be refined with more study, but let's start with what we have.

Water

Like any other livestock, goats should have constant access to fresh water. But all feeds contain water, too. Water is vital to life, of course, but it's also important in feed formulations because the quantity of water in various plants affects their place in the ration. Dry grain, for example, might contain 8 to 10 percent water. Green growing plants might contain 70 to 80 percent water. An animal fed succulent plants ingests an enormous amount of water along with fiber in order to get at the nutrients available in the plants. In human terms, it would be the equivalent of drinking eight glasses of water with a meal. There wouldn't be much room left for the actual food. Goats will control their own water intake if they can, and they generally wait until after they have filled up on dry forage before visiting the water bucket.

Carbohydrates

Of the plants' dry matter, about 75 percent is carbohydrates, the chief source of heat and energy. These carbohydrates include sugars, starch, cellulose, and other compounds.

The sugars and starches are easily digested and have high feed value. Cellulose, lignin, and certain other carbohydrates, however, are digested only with great difficulty, and it takes energy to digest them: their feed value is correspondingly lower (this is one reason goat raisers prefer "fine-stemmed, leafy green hay." The fine stems mean less lignin and hard-to-digest materials).

If you buy feed, the feed tag on the sack will have the carbohydrates divided into two classes: crude fiber (or just plain fiber) and nitrogen-free extract. Nitrogen-free extract is the more soluble part of the carbohydrates and includes starch, sugars, and the more soluble portions of the pentosans and other complex carbohydrates. It also includes lactic acid (found in milk) and acetic acid (in silage). Oddly enough, nitrogen-free extract also includes lignin, which has a decidedly lower feeding value than cellulose.

Fats

Feed tags also list "fat," which actually includes fats and oils. They're the same except that fats are solid at ordinary temperatures while oils are liquid. In grains and seeds, fat is true fat. In hays and grasses, much fat consists of other substances. Many of these are vital for life, including cholesterol, ergosterol (which can form vitamin D), and carotene (which animals can convert into vitamin A). Note that these are all fats from plant, not animal, sources.

Proteins

Proteins and other nitrogenous compounds are of outstanding importance in stock feeding. Proteins are exceedingly complex, each molecule containing thousands of atoms. There are many kinds of proteins, some more valuable than others (livestock feeders speak of the "quality" of protein). All are made up of amino acids, of which at least 24 have been identified, and protein must be broken down into amino acids before it can be absorbed and utilized by the body. Because they can combine like letters of the alphabet, there could be as many proteins as there are words in the dictionary.

The protein in plants is concentrated in the rapidly growing parts (the leaves) and the reproductive parts (the fruits or seeds). In animals, protein makes up most of the protoplasm in living cells and the cell walls, so it's important for muscles, internal organs, skin, wool or hair, and feathers or horns, and it's an important part of the skeleton.

Protein, or crude protein, includes all the nitrogenous compounds in feeds. It's of extreme importance to the animal caretaker, and it's obviously essential for life, but needs vary among classes of animals. Protein requirements are higher for young and growing animals, reproduction, and lactation. And because protein is the most expensive portion of livestock feed, you won't want to offer more than necessary.

Minerals

Ash indicates the mineral matter of the ingredients. Minerals in plants come from the soil, but the mineral content of animals is higher than that of plants. Calcium and phosphorus are particularly important, since they are the chief minerals in bone and in the body. The body contains about twice as much calcium as phosphorus, and the proper balance is important.

Other minerals are needed in trace amounts, but they are nevertheless vital. Iodine, for example, prevents goiter; iron is important for hemoglobin, which carries oxygen to the blood. Copper, which is a violent poison, is also a necessity in trace amounts. A lack of iron, copper, or cobalt can result in nutritional anemia, and a zinc deficiency results in crusty patches around the nose. We'll come back to minerals later.

Total Digestible Nutrients

Net energy values of livestock feeds are expressed in therms instead of calories. Since a therm is the amount of heat required to raise the temperature of 1,000 kilograms of water 1°C, one therm is equal to 1,000,000 calories.

Nutrients are constantly being oxidized in tissues to provide heat and energy. This oxidation maintains body heat and powers all muscular movements. Since the digestion of roughages requires more energy, it follows that 1 pound (0.45 kg) of total digestible nutrients, or TDN, in roughages will be worth less than 1 pound of TDN in concentrates, which will not use up so much of their energy just being digested. "Total digestible nutrient" refers to all the digestible organic nutrients: protein, fiber, nitrogen-free extract, and fat (note that fat's energy value for animals is approximately 2.25 times that of protein or carbohydrates).

"Digestible," of course, refers to nutrients that can be assimilated and used by the body. For this reason, protein or crude protein is different from digestible protein. Digestible nutrients are determined in the laboratory by carefully measuring the amount of feed consumed and analyzing its content, and then analyzing the waste products (animal feces are largely undigested food, in contrast to human feces, which have a larger proportion of spent cells and other true "waste").

Vitamins

Another important consideration in feeding is vitamins. Vitamins were largely unknown before 1911, and there is still more to learn about them. But as of now, the only two of any consequence to goats are vitamins A and D.

Vitamin A

Vitamin A is of prime importance to dairy goats because it's necessary for growth, reproduction, and milk production. It is of less importance in maintenance rations. Vitamin A is synthesized by goats that receive carotene in their diets; the chief sources of carotene are yellow corn and leafy green hay. Common symptoms of vitamin A deficiency are poor growth, scours, head colds and nasal discharge, respiratory diseases including pneumonia, and blindness. A severe lack of vitamin A prevents reproduction or produces weak (or dead) young at birth.

Vitamin D

The other important vitamin for goats is vitamin D. As with other animals, lack of this vitamin causes rickets, weak skeleton, impaired joints, and poor teeth. Vitamin D is necessary to enable the body to make proper use of calcium and phosphorus. The best and chief source is sunshine, but it is also available in sun-cured hay.

Other Vitamins

The B-complex vitamins are manufactured in the rumen, so the feeder has no concern with them directly. Vitamin E seems to have no special application to goats except in its association with selenium. Vitamin C is synthesized (only humans, monkeys, and guinea pigs lack the ability to manufacture vitamin C). Vitamin K is also synthesized.

Formulating a Goat Ration

With this very brief background of a goat's nutritional requirements, we can begin to formulate a ration. Since we now know how rumination works, and how important roughage is to that process, we'll begin with roughage. We'll then consider grains before developing some sample rations.

Roughage

Roughage can be green, growing plants, including grasses, clovers, and the trees and shrubs goats eat. It can be plants in dried form, called hay. There are two types of hay: legume hay made from alfalfa or clover; and carbonaceous hays made from timothy, brome, or other grasses. Corn stover (dry cornstalks), silage (fermented corn plants or hay plants), comfrey, sunflower and Jerusalem artichoke stems and leaves, and root crops such as mangel beets, Jerusalem artichokes, carrots, and turnips are also considered roughages.

Green Forages

Green forages are rich in most vitamins, except vitamins D and B_{12}. But if the animal is grazing, it is getting vitamin D with sunshine, and ruminants can synthesize B_{12}. Rapidly growing grass is also rich in protein.

However, because of the high water content of succulent green feed (and roots, too), these are low in minerals. The lack of minerals, combined with the high water content

(an animal could drown before it got enough nutrients from really lush grass), means that such forage does not constitute an adequate diet by itself. And lush forages can cause bloat.

This is not to say that green feed is not desirable but only that it must be understood and properly managed.

Confinement Feeding. There are two ways to use green forages. One is the familiar pasture, where the animals "harvest" their own feed. The other is confinement feeding, where the caretaker does the harvesting and brings the feed to the animals. In this case, the goats are usually restricted to a loafing barn and a

---

Clover makes excellent roughage, both as growing plants and as hay, but clover — as well as alfalfa (page 84) — should be harvested before the blooms open.

clover

relatively small, and therefore mostly unvegetated, exercise yard.

Confinement feeding has been popular among people with a few goats for many reasons. Often land is limited. Goats are notoriously difficult to fence, and fencing large areas can be expensive. Unlike cattle or sheep, goats dislike grazing in inclement weather, and goats, more than cattle, are easy prey for stray dogs and other predators. Owners of a few goats are less likely than large farmers to be around during the day to keep an eye on things or to have the time to manage rotational grazing. Animals on pasture waste a lot of feed by trampling and selective grazing. Pasturing allows the caretaker less control over what the animal eats, including toxic plants and those that can affect the flavor of milk. And because most home dairy owners with just a few animals (and outside jobs) purchase their hay and grain, feeding in confinement makes sense.

Farmers who choose not to rotationally graze but have sufficient acreage and equipment usually feed green chop — crops such as alfalfa or oats that are cut and chopped with tractor-drawn equipment and blown into special wagons. Some are feeders on wheels, while others require augering the feed out of the wagon and into feed bunks. Even if you have just a few goats, you can cut green feed with a scythe, string trimmer, or similar tool, or even a scissors or butcher knife, and carry it to the animals by the armload or in a barrow or cart.

Consider planting forage crops in your garden. Kale, chard, carrots, and others are popular, but oats, rye, and alfalfa grow well, too.

Grass Clippings

Goats are not lawn mowers, and lawns are not pastures. And the better your lawn — from a homeowner's perspective, that is — the worse it is for goats. Goats like variety and dislike eating off the ground, perhaps as an instinctive defense against parasite infestation. Lawns strike out on both counts.

Lawn grasses have been developed to withstand foot traffic and frequent mowing, and for color, among other things, not for animal nutrition. They are high in moisture and very low in fiber and nutrients. They tend to pack tightly (both in the feeder and in the rumen), and they heat up and mold quickly. They take up rumen space that could better be used by more nutritious feeds. And because of their poor nutritive quality, you'd have to feed more expensive concentrates to provide a balanced diet.

If you have a "beautiful" lawn, it's probably because you applied nitrogen fertilizer, which can mean toxic levels of nitrates in the grass. And naturally, if herbicides have been used, that grass is totally off-limits to goats.

On the other hand, some of us don't have or even want picture-perfect lawns like that. Ours are a combination of grasses and weeds, perhaps including clovers and chicory, and, of course, plenty of dandelions. We never fertilize the lawn and certainly never spray it.

In this case, one way to utilize that grass is to mow it without a grass catcher. Let it lie until it's dry. This might take a few hours on a hot, dry day, or it might take a day or two. This is basically lawn hay. It can be fed to the goats in limited quantities, or if it's dry enough, stored in plastic bags or a haymow. If you choose to feed fresh grass clippings, provide only as much as the goats will clean up in perhaps half an hour; remove the leftovers before they heat up. And provide other feeds as well to lessen the packing effect of the grass clippings in the rumen. (See also Making and Feeding Silage, page 266.)

Pasture

Pasturing goats presents both opportunities and problems. It saves labor, because the goats harvest the forage and spread manure, and it also seems like an obvious way to reduce feed costs — the major expense for the home dairy — especially if you have suitable land available. But before you take these economies for granted, check into the cost of fencing! (*Note:* One acre [0.4 ha] of land will require at least 825 feet [250 m] of fencing, and more if it's not square.)

Also, take an inventory of what's growing in your proposed pasture. Goats prefer variety, and woody plants rate high with them. Watch for poisonous plants (see page 86).

In most cases, animals on pasture trample and otherwise waste more than they eat. And goats are picky, taking a bite here, a nibble there. With their almost prehensile lips, they can select exactly what they want, which is not necessarily what you want for them. They don't graze down to the ground as sheep and cattle do, making it difficult to know just how much nutrition a goat is getting from browse. Laboratory tests of plants determine the food value of the entire plant, but goats generally select just the part they want.

Good Pasture Management. Good pasture management involves providing the best pasture in each season with a stocking rate that is compatible with good renewal of the vegetation and the best sustainability of forages and browse. Your county or parish Extension agent can help with information and recommendations based on your soil and forage types.

The seasonal factor is important, because different plants grow best under different conditions. Renewability and sustainability are important because if left on their own, without rotation, goats (and sheep and cattle) will eat the plants they favor, perhaps killing them out. The less desirable plants will take over.

Seasonality also involves nutrition. A dramatic example can be seen in the lush pastures of spring becoming sparse and brown after a summer's hot, dry spell. On poor pasture a goat will not produce milk, and her

A well-fenced area with a variety of plants provides a fine pasture that makes raising goats easy and economical.

health might even be endangered without supplemental feed.

Ideal pastures are soil tested and properly fertilized; planted to specific desirable species; and managed to avoid overgrazing and to encourage overall productivity, preferably including rotational grazing.

Rotational Grazing. The simplest way to utilize pastureland is to fence it in, turn in the goats, and let 'em at it. But much as most goat owners cherish simplicity, it isn't always the best way.

Any grazing or browsing animal introduced to a pasture will select its favorites first. This isn't quite like a child who, given free rein at the dinner table, starts with the dessert: it's worse. The problem is damage to the pasture. The preferred parts of the preferred plants are eaten with relish, usually faster than they can regrow. Meanwhile, other plants that might be just as nutritious are left to grow old and coarse. Overgrazing can eradicate the most favored plants, while the others become useless.

With rotational grazing, the pasture is divided into paddocks, or smaller pastures. How many depends on several factors, including how much time and money you want to spend on fencing. Some people use only a few paddocks, while others have twenty paddocks or more.

The animals are allowed into one section, where they first seek out their favorite foods. But because there aren't as many of those in the smaller space, less favored plants are also consumed.

Then after a few days — or more, or less, depending on the degree to which the plants have been eaten — the livestock are moved to a new section. The feeding process starts all over again, while the previous pasture gets a chance to regrow.

Movable fencing makes this much more practical than it would be with permanent fencing, and the additional forage produced on the same amount of land can make such a system worthwhile. There are other benefits, such as increasing the sustainability of the land. One drawback, in addition to the need for additional fencing, is providing drinking water in each section. A good solution for goats is to design the fence layout so the animals can return to the barn to drink or if it rains.

Managed rotational grazing provides many benefits, and if it interests you, you'll want to learn more about it. *Stockman Grass Farmer* and *Graze* are publications devoted to this system.

Weeds and Poisonous Plants. A variety of weeds in a pasture can be both attractive and beneficial to goats. Some common ones include chicory, daisies, dandelions, nettles, plantain, thistles, and yarrow. And some plants that humans try to avoid are safe for goats, including nettle and poison ivy. Goats also love some serious pest weeds, such as multiflora rose, which are not mentioned in the scientific literature as causing any health problems for goats.

GOATS ON GRASS

Studies show significantly higher levels of beneficial omega-3 fats and conjugated linoleic acid (both related to a decreased risk of cardiovascular disease in humans) in meat and milk from grass-fed cattle as compared to grain-fed cattle. One can hypothesize that milk and meat from grass-fed goats would also contain higher levels of these nutrients, but we'll have to wait for the studies to know for sure.

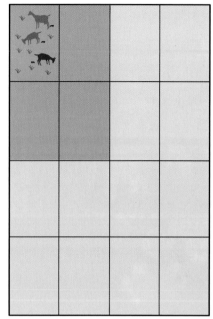

Managed rotational (or intensive) grazing involves moving animals from one paddock, or small pasture area, to another as the forage plants are consumed. Pastures can be subdivided with permanent fencing, or temporary fencing can be used and moved as the animals need new forage. Animals should be moved before forage is shorter than 6 inches. (Research shows that 90 percent of parasite larvae are found on the lower 4 inches of grass and 75 percent are on the lower 2 inches.)

Sorrel and dock are often considered valuable feed weeds, but these are well-known oxalate accumulators, which can be toxic. Spinach has been associated with oxalate toxicity of goats in Australia, as has amaranth (pigweed) in Mexico. Grazing goats are unlikely to consume enough of these to cause problems, but it's not a good idea to cut armloads for goats in confinement. (Oxalate toxicity also results from chemicals such as ethylene glycol — automotive antifreeze. Because of its toxicity, it is just as important to keep antifreeze away from your goats as it is to keep it away from your pets and your children.)

Poisonous plants aren't a common problem with goats unless they are desperate for food. While you are not going to let them get to that point, you should certainly be aware of what's growing in your pastures and their possible effects on goats. Arm yourself with a good plant-identification book for your region, and ask your county Extension agent what problem plants you should be watching for. Goats' browsing habit of taking a bite here and a nip there tends to protect them from consuming too much of most dangerous plants. But be on the lookout for wilted wild cherry, oak leaves, rhubarb leaves, milkweed, hemlock, mountain laurel, and locoweed, all of which are poisonous to goats. Bracken fern, which is toxic to cattle and sheep, seldom causes problems in goats.

88

GOOD WEEDS FOR GOATS

chicory

daisy

dandelion

multiflora rose

nettle

plantain

thistle

yarrow

BAD WEEDS FOR GOATS

bracken fern

dock

hemlock

locoweed

milkweed

mountain laurel

oak leaf

rhubarb

sorrel

wilted wild cherry

Nitrate Poisoning

Goats convert nitrates to nitrites, as do plants. Nitrate poisoning is caused not by nitrates but by their overaccumulation in the goat's system.

Some plants normally considered good feed can undergo chemical changes due to weather. Sudan grass, johnsongrass, pigweed, lamb's-quarters, alfalfa, corn, and oats can accumulate toxic amounts of nitrates if they undergo rapid growth after a dry spell. Low temperatures, decreased light, and, of course, heavily fertilized soils can also adversely affect these plants, causing nitrate accumulation.

Nitrate poisoning can also be caused by water contaminated with animal wastes or runoff from fertilized fields or by eating fertilizers. Several cases of nitrate poisoning in goats have been blamed on water provided in galvanized containers.

Hay

Understanding hay is an important part of raising goats. But many people who are new to goats have no background with farm animals and no experience with hay. Some don't even know the difference between hay and

johnsongrass

amaranth (pigweed)

straw. Hay refers to grasses or legumes cut at an early stage of growth and sun-dried. Hay should be more or less bright green and is used as feed for equines and for ruminants such as goats. Straw is the dried leaves and stems of grasses grown for grain, such as oats, wheat, and barley, after the grain has been removed by combining. Straw is a golden color and is used primarily as bedding.

Not all hay is created equal. The first major division is between grass, or carbonaceous hay, and legume hay. Grass doesn't refer to lawns! Grass hay might be timothy, johnsongrass, orchard grass, or others grown specifically as

alfalfa

animal feed. Legume hay comes from alfalfa or the clovers.

The carbonaceous hays have less protein and less calcium than the legumes, and these deficiencies must be made up in the concentrate or grain ration.

Other hay plants can include barley (cut when the seed heads are in the immature, or "boot," stage), bird's-foot trefoil, Bermuda grass, lespedeza, marsh or prairie grasses, oat or wheat grasses, soybeans, or combinations of these. For the benefit of inexperienced farmers, it will be well to point out again that hay is made by cutting green growing plants and drying, or curing, them in the sun. Wheat, barley, oats, and soybeans, for example, can be cut when young, for hay. If allowed to mature, the nutriment goes into the grain, and the stems and leaves become yellow and have little food value; the plant that could have been hay becomes straw.

Hay can also be referred to as first, second, or third crop or cutting (and more in some climates). First cutting generally has coarse stems and less total digestible nutrients when compared with later cuttings. The stage of maturity when cutting hay is crucial, especially for legumes. Alfalfa and clover should just be starting to blossom; after that the nutritive value decreases rapidly. Some hayfields include weeds of various kinds, which must be considered when trying to determine the feed and dollar value of the hay.

Hay Quality. But all of this can be overshadowed by how the hay is cured in the field.

Johnsongrass, amaranth (pigweed), and alfalfa, normally good feed for goats, can cause nitrate poisoning in certain situations.

If it rains while the hay is drying, the feed value can be diminished or even destroyed. If the hay is baled while too wet (either from rain or inadequate curing), mold is a problem; if it's too dry, the leaves containing most of the nutrition are likely to be shattered and lost in the baling process. Some hay is conditioned by passing through rollers that crush the stems, allowing for more rapid and uniform drying of stems and leaves. In broad terms, the faster the drying, the better the quality.

The importance of quality hay can be illustrated by the fact that good alfalfa can have as much as 40 milligrams of carotene per pound, while alfalfa that is bleached and otherwise of poor quality can have as little as 4 milligrams per pound. Poor hay may be difficult to distinguish from some straw. Straw contains much fiber, especially lignin, and is mixed into total mixed rations of dairy and beef cattle. It doesn't add nutrients to the diet, but the fiber aids in digestion. Since goats choose to add fiber to their diets in other ways, they will rarely eat straw.

Alfalfa or clover hay is considered the ideal for goats because of the high protein content. Good alfalfa has about 13 percent protein; timothy and brome are usually closer to 5 percent. Alfalfa and clover are also rich in calcium, the most important mineral.

Experienced producers and buyers can tell a great deal about hay by its color, texture, aroma, and general appearance. But your county Extension agent can also tell you where and how to have hay tested so you know exactly what you're getting. Forage testing can help you determine what kind of grain ration to supplement hay with, for both performance and economy. If your hay is very good, you might use a grain mix with less protein, which can save money. If the hay has less protein than the average, you'll want to adjust your grain mix to compensate for that. What's just as important is that hay testing can help determine a fair price.

Hay Prices. Hay prices vary considerably from place to place and year to year. It's not unheard of for hay in the Midwest to go for $2 a bale and in California for $12 a bale (but bales are bigger in California). A mature goat will require anywhere from 3 to 10 pounds (1.5 to 4.5 kg) of hay per day, depending on type, quality, waste, and other factors.

Weather is a big determinant. If the rains are timely and sufficient, farmers have plenty of hay, and the price goes down. In times of drought, however, farmers might not have enough to feed their own animals, and the price goes up. Alfalfa winterkill, the alternate thawing and freezing that forces the crowns out of the ground, can also cause prices to shoot up.

The most elusive factor for most goat owners is quality. Pure alfalfa, properly cut in the early bud stage, conditioned, and properly cured, generally costs more than grass hay. Fine-stemmed hay is worth more than coarse-stemmed hay. The first cutting, or first crop, is usually cheaper than second or third, but this distinction may be meaningless in an area where in a given year, there is no second or third cutting.

CAUTION!

At certain stages of growth, white clover can be toxic to goats. Be sure there is plenty of other browse available so they aren't forced to eat what they shouldn't.

WEIGHTS OF SOME COMMON GOAT FEEDS

Ingredient	Weight in Pounds per Quart
Barley, whole	1.5
Buckwheat, whole	1.4
Corn, whole	1.7
Linseed meal	0.9
Molasses	3.0
Oats	1.0
Soybeans	1.8
Sunflower seeds	1.5
Wheat, whole	1.9
Wheat bran	0.5

But then, you might not need or even want the very best quality hay. Dry does and bucks might get different hay from what the milkers get, and coarse hay provides more body heat than fine during the cold days of winter.

Grains: The Concentrate Ration

While roughages are the most important part of the diet of a ruminant, they alone don't provide all the needed vitamins and minerals, nor do they provide sufficient energy. Alfalfa hay has about 40 therms per 100 pounds (45 kg); corn and barley each have twice that. Especially if you feed carbonaceous hays, which have only 5 percent protein, you must provide additional protein and calcium. Hays do not provide sufficient phosphorus. These missing elements are provided in the concentrate ration.

The concentrate ration is often called the grain ration, but this term can be misleading. Here's why: For lactating animals, the protein content of the concentrate ration should be about 16 percent if the roughage is a good legume. With less protein in hay, more must

be added to the concentrate. For dry does, 12 percent protein is sufficient. Corn has about 9 percent protein and only 6.7 percent digestible protein. Oats have about 13 percent protein and 9.4 percent digestible protein. Therefore, a mixture of equal parts of corn and oats would contain 11 percent protein or about 8 percent digestible protein.

Clearly, these grains alone will not meet the demands of the growing or milking animal. Therefore, protein supplements in the form of soybean-oil meal (sometimes listed on feed tags as SOM), or linseed, or cottonseed-oil meal must be added. Many goat owners try to avoid cottonseed-oil meal because of the massive amounts of pesticides used on that crop.

Milking animals also require more salt than is needed for animals on maintenance rations. It is usually added at the rate of 1 pound (0.45 kg) per 100 pounds (45 kg) of feed.

Because of the need for bulk in the diet of a ruminant, a concentrate ration should not weigh more than 1 pound per quart (about 1 L). Bran is most commonly used for bulk. (Beet pulp is sometimes used for does, but extended feeding of beet pulp to bucks can cause urinary calculi.) The weight of grain varies with quality, which is often determined by the weather during the growing season, but this chart shows some normal averages.

Molasses: An Important Extra

Finally, since goats generally shun finely ground feed such as that normally fed to cows, the grains should be crimped or cracked or even whole, rather than ground to flour. Cows do not digest whole grains well. Whole corn goes in one end and out the other. Goats seem to have better powers of digestion and actually pick out the large morsels to eat first.

This is fine for the goat owner who wants to mix a ration rather than buy bagged feed, because it eliminates the bother and expense of grinding. But it also means the fine ingredients — salt, bran, oil meal, and minerals — can't be mixed into the grain. They sift to the bottom and the goats won't consume them. To overcome this, most goat feeds contain cane molasses. In addition to binding the ingredients, molasses makes the feed less dusty, it's an important source of iron and other important minerals, it increases the palatability of the feed, and does fed ample molasses during gestation are less likely to encounter ketosis (see chapter 8). Molasses contains about 3 percent protein, but none of it is digestible.

A popular alternative to loose salts and minerals is to add a pelleted protein to the mix. It often comes as a complete mineral and contains as much as 36 percent protein. A nutritionist can help determine the other components that will make a properly balanced ration for milking goats, dries, bucks, or kids. Even with protein pellets, molasses can still be a valuable added nutrient.

There is some evidence, at least in dairy cows, that excess molasses interferes with the digestibility of other feeds. The digestive processes attack the more easily assimilated sugars in molasses to the detriment of other feedstuffs. This is one reason some authorities advise against giving goats horse feeds, which are high in molasses (many also contain copper in amounts toxic to goats). Feeds with more than 5 to 6 percent molasses should be avoided.

Creating a Balanced Ration

At last we're ready to formulate a ration. The main tool we'll use will be a list of the protein content of the common goat feeds (see pages 96–97). The idea is to combine the various ingredients you have available in such a way that the combination will contain the desired amount of protein or, more accurately, digestible protein. But since protein is only one element of feed value, we must also keep in mind the minerals, vitamins, fiber, and palatability.

Just as importantly, any given ration depends on locally available ingredients and their comparative prices. Suggested rations almost invariably have to be adjusted. Unless the feeder knows what to look for, the carefully formulated suggestions will be thrown out of balance by indiscriminate substitutions. Even if you find a suggested formula you and your goats like, be sure that you or the nutritionist at your local mill check it out to be sure it is the protein level you intend.

Likewise, even the person who feeds commercial rations can destroy the balance by haphazardly adding "treats" or by making use of available grains in addition to the commercial feed. You can no more prepare a balanced diet by adding a handful of this to a scoop of that than you could expect to bake a cake by using the same method.

To complicate matters, the feed value of hay and grain varies from place to place and year to year, being affected by soil, climate, and other factors. For example, grains grown on the Pacific Coast are lower in protein than those grown elsewhere; also, the nutrients in hay harvested at the proper stage of development and well cured will differ dramatically from the nutrients in hay that is cut too late and leached or spoiled by improper curing (reminder: moldy hay should never be fed).

IS IT WORTH THE TROUBLE?

Is it worth going through all this, when it's so easy to buy scientifically formulated feed in convenient 50-pound bags? For most people, the answer is a resounding "No!" Initially, many people are shocked by the cost of commercially prepared bagged feeds. After all, they reason, if you can buy 100 pounds of corn for $5 or $6, why pay $15 or more for 100 pounds of feed mixed specifically for goats? Their question is answered when they stop to consider the hassle and expense of buying, storing, handling — and mixing — the individual ingredients.

On the other hand, some people want organic feeds that have been grown without chemical fertilizers, herbicides, or other pesticides and processed without antibiotics, preservatives, or medications. Today, we can add genetically modified plants to this list. It is getting easier to find premixed organic feeds, but you may have better luck finding organically grown grains and mixing your own.

Then there are people who already grow hay and grain, or who will, when they get some goats. Naturally, they'd rather feed that to the goats than buy a commercial mix.

These rations are provided for those dedicated do-it-yourselfers. But studying them will help you get a better idea of what a goat diet should be like, even if you buy your feed ready-mixed.

AVERAGE COMPOSITION OF SELECTED GOAT FEEDS

Feed	% Crude Protein	% Digestible Protein	% Fat	% Fiber
Dried Forages				
Alfalfa hay	15.3	10.9	1.9	28.6
Bermuda grass	7.1	3.6	1.8	25.9
Birdsfoot trefoil	14.2	9.8	2.1	27.0
Brome	10.4	5.3	2.1	28.2
Johnsongrass	6.5	2.9	2.0	30.5
Mixed grass	7.0	3.5	2.5	30.9
Red clover	12.0	7.2	2.5	27.1
Soybean, early bloom	16.7	12.0	3.3	20.6
Timothy, early bloom	7.6	4.2	2.3	30.1
Succulents (fed green)				
Alfalfa, green, early bloom	4.6	3.6	0.7	5.8
Bermuda grass, pasture	2.8	2.0	0.5	6.4
Cabbage	1.4	1.1	0.2	0.9
Carrot roots	1.2	0.9	0.2	1.1
Kale	2.4	1.9	0.5	1.6
Kohlrabi	2.0	1.5	0.1	1.3
Mangel beets	1.3	0.9	0.1	0.8
Parsnips	1.7	1.2	0.4	1.3
Potatoes	2.2	1.3	0.1	0.4
Pumpkins (with seeds)	1.0	1.3	1.0	1.6
Rutabagas	1.3	1.0	0.2	1.4
Sunflowers (entire plant)	1.4	0.8	0.7	5.2
Tomatoes (fruit)	0.9	0.6	0.4	0.6
Turnips	1.3	0.9	0.2	1.1
Grain Mix Ingredients				
Barley	12.7	10.0	1.9	5.4
Bone meal, steamed	7.5	—	1.2	1.5
Buckwheat	10.3	7.4	2.3	10.7
Cane molasses	3.0	—	—	—
Corn, #2 dent	8.7	6.7	3.9	2.0
Field peas	23.4	20.1	1.2	6.1
Linseed meal	35.1	30.5	4.5	9.0
Oats	12.0	9.4	4.6	11.0
Pumpkin seed	17.6	14.8	20.6	10.8
Rye	12.6	10.0	1.7	2.4
Soybeans	37.9	33.7	18.0	5.0
Sunflower seed, with hulls	16.8	13.9	25.9	29.0
Wheat (average)	13.2	11.1	1.9	2.6
Wheat bran	16.4	13.3	4.5	10.0

% Nitrogen-Free	% Mineral Matter	% Calcium	% Phosphorus Extract
36.7	8.0	1.47	0.24
48.7	7.0	0.37	0.19
41.9	6.0	1.60	0.20
39.9	8.2	0.42	0.19
43.7	7.5	0.87	0.26
43.1	6.5	0.48	0.21
40.3	6.4	1.28	0.20
37.8	9.6	1.29	0.34
44.3	4.7	0.41	0.21
9.3	2.1	0.53	0.07
12.2	3.1	0.14	0.05
4.4	0.7	0.05	0.03
8.2	1.2	0.05	0.04
5.5	1.8	0.19	0.06
4.3	1.3	0.08	0.07
6.0	1.0	0.02	0.02
11.9	1.3	0.06	0.08
17.4	1.1	0.01	0.05
5.2	0.9	—	0.04
7.2	1.0	0.05	0.03
7.9	1.7	0.29	0.04
3.3	0.5	0.01	0.03
5.8	0.9	0.06	0.02
66.6	2.8	0.06	0.40
3.2	82.1	30.14	14.53
62.8	1.9	0.09	0.31
61.7	8.6	0.66	0.08
69.2	1.2	0.02	0.27
57.0	3.0	0.17	0.50
36.7	5.7	0.41	0.85
58.6	4.0	0.09	0.33
4.1	1.9	—	—
70.9	1.9	0.10	0.33
24.5	4.6	0.25	0.59
18.8	3.1	0.17	0.52
69.9	1.9	0.04	0.39
53.1	6.1	0.13	1.29

Sample Ration Formulas

These rations, followed more or less faithfully, can be expected to produce good results. There will be minor variations because the feed value of grains depends in part on variety, weather, and the fertility of the soil that produced them. Even your choice of whole or steamrolled corn will make a difference; steamrolled corn is much more expensive but is significantly higher in digestible protein for a goat.

If certain ingredients are not available in your locale or others are more common and therefore less expensive, different grains can be substituted for one another by using the chart showing protein contents on pages 96–97.

Determining Pounds of Protein

You can determine the weight of protein in a given feed ingredient by multiplying the number of pounds of the ingredient by its percent of digestible protein. If you work in batches of 100 pounds, to figure the percent of protein in the whole ration, merely move the decimal point two places to the left.

As an example, let's look at a small homestead farm that produces its own grain. The previous year's corn crop was almost a total failure due to a wet spring, summer drought, and early frost. But other grains were available. Below is what the milking does were fed.

It should be noted that some rations you will find elsewhere work with crude protein rather than digestible protein. It may be easier to obtain figures on crude protein for locally grown feeds from your county Extension office, in which case all the ingredients should be calculated on the basis of crude protein.

RATION BALANCE CALCULATOR

Now that you have made it this far in the learning curve, it's only fair to mention that Langston University has a very good ration balancer and nutrition calculator based on the most recent research into goat nutrition. Visit *www2.luresext.edu/goats/research/rationbalancer.htm.*

Feed	Weight (Lbs)	% Crude Protein	% Digestible Protein	Protein (Lbs)
Barley	19	12.7	10.0	1.90
Buckwheat	5	10.0	7.4	0.37
Corn	10	9.0	6.7	0.67
Linseed meal	10	34.0	30.6	3.06
Molasses	10	3.0	0	0
Oats	20	13.0	9.4	1.88
Salt	1	0	0	0
Soybeans	20	37.9	33.7	6.74
Wheat bran	5	16.4	13.3	0.66
TOTAL	**100**			**15.28**

Do the math: Divide the pounds of protein by the total pound weight of the ration, then move the decimal two points to the right to find the percent protein.

SAMPLE RATIONS

Feed	Pounds (kg)
For a milking doe fed alfalfa hay (for a total of 12.6 percent digestible protein)	
Cane molasses	10 (4.5)
Corn	31 (14)
Linseed-oil meal	22 (10)
Oats	25 (11)
Salt	1 (0.5)
Wheat bran	11 (5)
or	
Barley	40 (18)
Cane molasses	10 (4.5)
Oats	28 (13)
Salt	1 (0.5)
Soybean-oil meal	11 (5)
Wheat bran	10 (4.5)
For a dry doe (for a total of 9.8 percent digestible protein)	
Corn	58 (26)
Oats	25 (11)
Salt	1 (0.5)
Soybean-oil meal	5 (2.25)
Wheat bran	11 (5)
For a milking doe fed nonlegume hay (for a total of 21.2 percent digestible protein)	
Cane molasses	10 (4.5)
Corn	11 (5)
Corn-gluten feed	30 (14)
Oats	10 (4.5)
Salt	1 (0.5)
Soybean-oil meal	24 (11)
Wheat bran	10 (4.5)
or	
Barley	25 (11)
Linseed-oil meal	15 (7)
Oats	20 (9)
Salt	1 (0.5)
Soybean-oil meal	25 (11)
Wheat bran	10 (4.5)
For a dry doe or buck (for a total of 10.1 percent digestible protein)	
Barley or wheat	51.5 (23.5)
Oats	35 (16)
Salt	1 (0.5)
Wheat bran	12.5 (5.5)

A Few Special Considerations

This leads us to another — perhaps the most important — reason every goat owner should have at least a basic knowledge of feed formulations. Goat keepers are notorious for dishing out treats or making use of "waste." These are both admirable pursuits, but they can cause trouble.

Watch the Balance. Assume that a goat receives 1 pound of a commercial 16 percent (crude) mixture. Maybe it costs the owner $16 per cwt, and he can get corn for half that, or he grew a little corn for the chickens and has some extra. Or the goat just seems to "like" corn! He decides to give the goat a pound of the regular ration and a pound of corn. So out of 100 pounds of feed, the goat gets 50 pounds of goat feed, which amounts to 8 pounds of protein, and 50 pounds of corn, which amounts to 4.5 pounds of protein. The total protein per 100 pounds of feed is 12.5 pounds or 12.5 percent. So the goat's protein intake has been reduced from 16 percent in the original feed mixture to 12.5 percent. That might be enough for the goat to maintain her own body but not to produce kids and milk.

The same thing happens when the animal is given garden "waste" or trimmings. Such fodder replaces roughage, not grain, but even then it can cause unbalancing of the diet because elements of hay, for instance, will be missing from most of the garden produce.

FEED-COST GUIDE

To be profitable, total feed costs should not exceed one-half the value of the milk produced.

Another caution is that your goats have a say in all this. Despite the rude claim that goats will eat anything, they are really picky eaters. You may have hit on a formula that fits all your goat's needs, but if your milker eats everything but the fuzzy little pieces of whole cottonseed, your carefully balanced ration goes down the tubes (this is another reason pelleted protein or molasses as a binder for fine minerals is so important in a ration). Try a small batch of 100 or 200 pounds (45 or 90 kg) — or whatever small amount the feed mill will mix—and make sure your goats like it before putting in a half ton for the winter.

This is not to say that rations can't be manipulated or that the goat breeder shouldn't make use of what's available or cheap. Just do it with a certain amount of knowledge and discretion.

Common Feeds for Small Farms

With this principle firmly in mind, let's examine some of the common feeds small farmers have available and show an interest in.

Soybeans. Soybeans deserve special mention because many people look at the price of the oil meals and wonder why the beans can't be fed whole. They can, with certain restrictions.

Soybeans contain what is called an antitrypsin factor. Trypsin is an enzyme in the pancreatic juice that helps produce more thorough decomposition of protein substances. The antitrypsin factor doesn't let the trypsin do its job, which means the extra protein in the soybeans is lost, not digested. The antitrypsin factor can be destroyed by cooking and isn't present in soybean-oil meal.

However, rumen organisms apparently inactivate the antitrypsin factor when raw

PLANT A GOAT GARDEN

Most goats are raised on small farms or homesteads where grain and hay are not produced. Such places can still grow a great deal of goat feed if the basic principles of feeding are followed.

"Grow milk in your garden" by planting sunflowers (the seeds are high in protein, and the goats will eat the entire plants), mangel beets, Jerusalem artichokes, pumpkins, comfrey, carrots, kale, and turnips, among others. In addition, you can utilize such "waste" as cull carrots, apples, and sweet corn husks and stalks in the goat yard. Treat them like pasture, using them to replace part of the grain ration but not all of it. Feed at least 1 pound (0.45 kg) of concentrates per head per day to milking animals.

soybeans are fed in small amounts. Current recommendations for dairy cattle are that the ration should not contain more than 20 percent raw soybeans. The same seems to work for goats. There is one major exception: do not feed raw soybeans if your feed contains urea! The result will probably be a dead goat. Here's why.

Urea. Urea is a nonprotein substance that can be converted to protein by ruminants. Urea is not recommended for goats, but many dairy feeds for cows contain it. So does liquid protein supplement (LPS), which some feed dealers will try to sell you when you ask for molasses (such dealers are not being dishonest; LPS is okay for cows but not goats, and often they simply don't know as much as you do about our caprine friends). Some people feed urea to goats because it's less expensive than the oil-meal protein supplements, but many goat keepers have reported breeding problems with animals fed urea. Toxicity can result from improperly mixed feed or when urea is fed along with a high-fiber diet that lacks readily digestible carbohydrate. The recommendation is to stay clear of urea altogether.

Gathering Weeds for Your Goats. Many people with more time than money and

a keen interest in nutrition are also avid collectors of weeds for their animals. For example, dandelion greens are extremely rich in vitamin A, and nettles are high in vitamins A and C. Goats relish these and other common weeds. It's just about impossible to imagine a real farmer on his knees gathering dandelion greens for his livestock, but many a goat farmer does and reaps healthier animals and lower feed bills.

Goats seem to enjoy variety more than most domestic animals, and no one plant has everything any animal needs for nutrition. Many goat keepers provide food from as many different plant sources as possible to enhance the possibility that their animals are getting the nutrition they need, naturally, without synthetic additives. They like grain mixtures of at least five or six different ingredients, and they like it very coarse.

Obviously, this isn't as efficient as modern agricultural methods. Farmers know that alfalfa is rich in protein and calcium, both important to dairy animals. A great deal of feed can be harvested from 1 acre (0.4 ha) of alfalfa, and alfalfa hay has become the norm. There are even herbicides to kill weeds in alfalfa to keep stands pure.

CAUTION!

Don't gather weeds from along roadsides where spraying is done.

But almost any weed in your garden has more cobalt than alfalfa. And cobalt is required by ruminants to provide the bacteria in the digestive tract with the raw material from which to synthesize vitamin B_{12}. Some, if not all, internal parasites rob their hosts of this vitamin.

Alfalfa and clover have little cobalt because lime in the soil depresses the uptake of this mineral, and lime is necessary for the growth and the calcium content of each. Agribusiness has found it more efficient to strive for high yields of alfalfa and then add the trace minerals to the concentrate ration. Homesteaders who don't mind gathering weeds can meet their animals' nutritional needs naturally and without the cash outlay required for commercial additives.

Organic farmers have known this for years, of course, but when their beliefs were confirmed by scientists in 1974, the idea was hailed as revolutionary. Researchers at the University of Minnesota compared the nutritive value and palatability of four grassy weeds and eight broadleaf weeds with alfalfa and oats as a feed for sheep, which have roughly the same requirements as goats.

Lamb's-quarter, ragweed, redroot pigweed, velvetleaf, and barnyard grass all were as digestible as alfalfa and more so than oat forage. All five weeds had more crude protein than oats, and four had as much as alfalfa. Eight were as palatable as oat forage.

GOATS LOVE TO TRIM THE TREES

Tree leaves and bark can be rich sources of minerals, brought from deep within the earth by tree roots. As far as goats' diets go, tree trimmings fall into the category of weeds. Some are great for goats; a few are dangerous. Here's a sampling.

- **Pine boughs.** Goats love them, but there have been reports that pine needles in large quantities have caused abortions, so caution is advised. Pine boughs are rich in vitamin C, although goats have no particular need for the vitamin, being able to manufacture it themselves.

- **Apple trees.** These are a great treat. By all means, feed the trimmings of apple trees when you prune your organic orchard.

- **Wild cherry.** Avoid altogether, since the wilted leaves are poisonous.

- **Oak leaves.** Avoid these also, as they're toxic for goats.

Comfrey. One particular plant deserves special attention, because so many people are interested in it and because it's controversial. That's comfrey, also known as boneset.

Several years ago, there was a rash of statements from county Extension agents and state departments of agriculture knocking comfrey. Some of their reasons for not growing it are practical — for large farmers, not

comfrey

homesteaders. And some of their information is just plain wrong.

It is true that a study conducted in Australia some years ago suggested that comfrey might be carcinogenic when fed excessively or over an extended period. That study didn't involve goats. And since then goats have consumed tons of comfrey, with no problems showing up in the scientific veterinary literature.

Even aside from that, comfrey should be in every goat owner's garden for at least limited use. Many goat and rabbit raisers swear by comfrey as a feed, a tonic, and medication for certain conditions such as scours. It's high in protein, ranking with alfalfa, although there is some question about the digestibility of the protein. But it is easier to grow and

harvest than alfalfa, using hand methods. It is an attractive plant that even can be used for borders or other decorative applications: grow goat food in your front yard or flower bed! It has tremendous yields because it begins growing early in spring and grows back quickly after cutting. And it's a perennial. It can be dried for hay, although that entails a lot of work because of the thick stems. It must be cured in small amounts on racks rather than left lying on the ground.

Minerals. No discussion of goat feeds would be complete without mentioning minerals. Most goat raisers supply loose minerals or mineralized salt blocks free choice and also add dairy minerals to the feed. There is some indication that this is unnecessary, expensive, and perhaps even dangerous to double up in the feed. Too much of a good thing can be as bad as too little. Goats that are well fed on plants and plant products from a variety of sources, grown on organically fertile soil, probably have little or no need for additional minerals.

But even when goats have a wide range of plants for forage, some still cannot fulfill their mineral needs. Plants grown in the

Goats are adept at selecting what is good for them to eat and what is not. Given the choice, they tend to seek out fresh growth and the most nutritious part of a plant species and stay away from the parts that could upset their digestive systems.

"goiter belt" (from the Great Lakes westward) are low in iodine, and iodized salt or kelp will be good insurance. Certain areas of Florida, Maine, New Hampshire, Michigan, New York, Wisconsin, and western Canada have soils deficient in cobalt. Parts of Florida are deficient in calcium.

Selenium deficiency can cause white-muscle disease (see chapter 8). Most soils in the central and eastern United States (and some in other areas) are deficient in selenium. Injecting does with selenium 15 to 30 days before they kid and kids at 2 to 4 weeks of age will prevent this serious disease. There's a relationship between selenium and vitamin E; both are usually administered at the same time. On the other hand, some soils, and the feeds grown on them, are high enough in selenium to cause poisoning.

Phosphorus is a vital ingredient of the chief protein in the nuclei of all body cells. It is also part of other proteins, such as the casein of milk. Therefore, it is of extreme importance to growing animals that are producing bone and muscle; to pregnant animals that must digest the nutrient needs for the growing fetus; and to lactating animals, which excrete great quantities of these minerals in their milk. Vitamin D is required to assimilate calcium and phosphorus. In addition, the ratio of calcium to phosphorus is critical. Most nutritionists agree on a 1.5:1.0 ratio. When legume forages are fed, the goat might need more phosphorus in the concentrate ration to maintain the balance; when carbonaceous forages are used, supplemental calcium might be required. Roughages, especially legumes, are high in calcium, and grains are high in phosphorus. If these crops are grown on soils rich in these minerals, the well-fed goat is likely to get enough of them.

A goat lacking phosphorus will show a lack of appetite, it will fail to grow or will drop in milk production if in lactation, and it may acquire a depraved appetite, such as eating dirt or gnawing on bones or wood (though many goats like to chew on wood even without a phosphorus deficiency). In extreme cases, stiffness of joints and fragile bones may result.

However, overfeeding calcium can be dangerous, too, especially for young animals. Lameness and bone problems can result later from excess calcium. It is also dangerous for bucks and wethers, which can develop urinary calculi.

Iron is 0.01 to 0.03 percent of the body and is vital for the role it plays in hemoglobin, which carries oxygen in the blood.

Copper requirements are about one-tenth those of iron, and in greater amounts copper is a deadly poison. Nutritional anemia can result from lack of iron, copper, or cobalt (this is different from pernicious anemia in humans). But it's very rare.

Other trace minerals are potassium, magnesium, zinc, and sulfur.

Most goat owners today prefer furnishing loose minerals free choice, rather than in block form, and find that their goats consume varying amounts depending on their needs. In this case there is no need to add minerals to the feed, too; as we have seen, it might not be a good idea.

DO GOATS KNOW WHAT THEY NEED?

Some livestock raisers believe animals can select the minerals they need. Theta Torbert of Maine told a writer for *Dairy Goat Journal* that her goats use a lot of iodine during the third month of pregnancy, magnesium just before they go on spring pasture, sulfur in August and January when they are growing their fur coats, phosphorus during the last 2 months of pregnancy, and a lot of potassium year-round. They don't seem to use calcium, but they do get dicalcium phosphate in their grain. If you want to check this out for yourself, furnish the loose minerals in separate mineral feeders.

The Science and Art of Feeding

Obviously, feeding is a science and an art. Goats are not "hayburners" or mere machines to be fueled haphazardly. You wouldn't burn kerosene in a high-powered sports car, and you can't get the full potential from a goat fed improperly. Remember these basic concepts:

- Feed your goats 1 pound (0.45 kg) of concentrate for maintenance and about 1 pound extra for each 2 pounds of milk produced, along with all the hay they will eat.
- The ration should come from as many different sources as possible and from fertile soil.
- Avoid sudden changes in feed, which result in overloading the rumen bacteria and microbes.
- Keep the mix very coarse for the best intake.
- Pay attention to protein levels, as well as vitamin and mineral content of the plants and grains you feed.
- Treat each animal as an individual, for they have different needs according to age, condition, production, and personal quirks.
- Some will do better on less, others will want more. That's the art, or part of it: "The eye of the master fattens the livestock."

7

GROOMING

Goats require a minimum of care, but that doesn't mean they require no care. Most of the basics of grooming, like brushing, clipping, and trimming hooves, yield the reward of a much healthier and happier goat. Other grooming basics, like disbudding or tattooing, may not win you an affectionate response at first, but both have benefits to you and the goat in terms of safety and peace of mind in the long run.

Hoof Care

The horny outside layer of the goat's hoof grows much like your fingernails and, like nails, must be trimmed periodically. Gross neglect of this duty can cripple the goat. But it's a simple job, and for the small herd it won't take more than a few minutes every couple of months. This makes it hard to understand why there are so many goats whose feet look like pointy-toed elf boots or even skis.

How Often?

How often you trim hooves depends on several factors, but at the very least you should check hooves every 2 to 3 months. Sometimes hooves grow faster than at other times, and there are differences among animals. Goats living on soft, spongy bedding will need more hoof attention than goats that clamber on rocks, which is how wild goats' hooves are worn down. One goat book claims that if a good-sized rock is placed in the goat pen, the animals will stand on it and keep their hooves worn down. It sounds logical, but everyone I know who has tried it says it doesn't work.

What Tools?

There are several methods of trimming, requiring different tools. The simplest trimming tool is a good sharp jackknife. Some people prefer a linoleum or roofing knife, and others swear by the pruning shears, the same kind you use for your roses. However, I don't think anyone who has used an honest-to-goodness goat-hoof trimmer would ever want to go back to the more primitive tools (some catalogs selling sheep equipment call them "foot rot shears," a rather ugly and misleading name for the same tool).

For light trimming and finishing, many people like a Surform, a small woodworker's plane with blades much like a vegetable grater. A stiff brush for cleaning and heavy gloves are also useful.

Goats love to stay clean and as a rule will stay away from wet, mud, and manure if given the option. Still, they require human intervention for some of their other grooming needs.

NATURAL HOOF MAINTENANCE

In nature, goats' hooves are worn down when they climb on rocks. If goats live on bedding or soft ground, hoof trimming becomes the job of the caretaker. Increase the time between hoof trimmings by placing a dozen or so rough cinder blocks in the yard or pasture, with holes turned to the side, obviously. The goats will love to climb on them, and the abrasive surface of the blocks will help minimize your trimming duties.

How to Trim Hooves

You'll want to know what a good hoof looks like before you start trimming. A young kid has the ideal hoof. When she stands, the bottom of the hoof is parallel with the part where the hair starts growing at the bottom of the pastern. Many older goats actually have growth lines that act as a handy template.

Unless you have docile goats or are rather robust yourself, hoof trimming is easier if you have a helper or if your milking stand is of a type that allows you to lock the goat in and still have room to get at all four feet (see chapter 13 for more information on milking stands).

With or without a milking stand, some goat owners work from the side and move around the animal as they finish each hoof. Others — the longer-legged ones — straddle the goat at the hips and then the shoulders,

Surform

pruning shears   goat-hoof trimmers

using their knees to control the goat's sideways movement. Still others tie her in a corner and press her against the wall opposite the hooves they are trimming. The object is to keep her from sidling away while you are trying to trim. No matter what method you use, be sure you have good lighting, and don't hurt the goat with excessive twisting or pulling on the joints. Some goats don't seem to mind such acrobatics; other will protest rather violently. It doesn't really matter which foot goes first, either, but we'll start from the back for this lesson.

1. **With the goat secured, lift one back leg so the hock is about hip high and the toe is turned up so you can work on it.** Keep a firm grip on the ankle, and be exceedingly careful with the knife or other tool so that a protesting kick or jump won't injure either of you. This is one reason many people prefer the shears, which are much safer than a knife. In either case, it's a good idea to wear sturdy gloves.

1

HELPFUL HINT

Dry hooves can be very hard. They're easier to trim after the goat has been walking in wet grass. Shears, which offer leverage, cut hard hooves more easily than a knife.

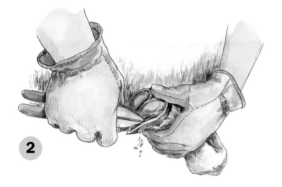

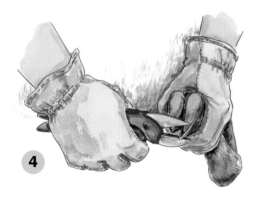

2. With the point of the tool or a stiff brush, **clean out all the manure and dirt embedded in the hoof.** If the hoof has not been trimmed for some time, it will have curled underneath the foot and can contain quite a lot of crud.

3. **Next, carefully cut off any hoof material that has overgrown the fleshy part of the foot.**

4. **Trim off excess toe.** The point of the hoof often wears down less than the sides and will require extra trimming. Heels seldom need trimming, but check them and cut very carefully, if necessary. You can cut quite safely until the white portions within the hoof walls look pinkish. Plane off only the horny hoof material, not the fleshy part of the foot! You'll have occasional slips and miscalculations, and even a tiny nick will generate a lot of blood, so it's a good idea to keep a container of blood-stop powder on hand.

5. The final step, which is optional, is **smoothing the hoof with a woodworker's Surform.** The portion near the toe invariably needs more of this finishing work than the heel.

6. **Let the goat stand on the hoof,** and see how it looks. A goat with good hooves stands squarely.

7. **Do the other hind hoof** the same way.

8. If your back has had enough strain at this point, you can **squat down beside a front leg,** bring the foot up over your knee, and trim as you did each hind hoof.

In extremely bad cases, where the goat looks as if it's wearing pointed elf's shoes, it may take several trimmings to get them back in shape. In such cases, it's better not to cut too much at once. For one thing, the blood supply grows farther out into the toe, so getting the foot back to the right length can include a good deal of blood if you try to do it all at once. For another thing, if the hooves are really bad, the goat isn't likely to stand around patiently while you finish. Take off as much as you can, and come back a day or two later. With feet that are in really bad condition, it might have to be as long as a month or two later.

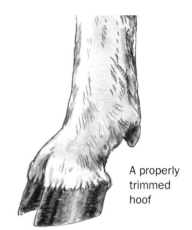

A properly trimmed hoof

Don't forget: bucks have hooves, too! These poor fellows are more likely to be neglected than the girls are, but of course they shouldn't be.

Disbudding

The other major grooming duty for goat owners is disbudding. Much has been written about the advantages and disadvantages of horns, and even more has been said around goat barns. Here's a summary of the arguments on both sides:

FOR HORNS	AGAINST HORNS
• Horns are protection against dogs and other predators. • Horns serve as a "radiator" to help cool the animal (Angora goats should always keep their horns, and any other breed used often for carting or packing should probably have them, too). • Horns are beautiful and natural, and a goat doesn't look like a goat without them. • Disbudding is ghastly.	• Horns are dangerous to other goats and to people, especially children. • It's impossible to build a decent manger that will accommodate a nice set of horns. • Horns regularly get caught in certain types of fences. • Horns are a disqualification on dairy breed show animals. • Horns aren't really much protection against dogs — look to better facilities instead or perhaps a guard dog, donkey, or llama. • Owners are tempted to use horns as built-in handles, which causes goats to learn bad habits like butting or poking in retaliation.

One thing everyone agrees on: disbudding is much better, and easier, than dehorning. Disbudding involves cauterizing the blood and nerve supply to the horn bud on a very young animal, before the horns really start to grow. Dehorning, on the other hand, is the surgical removal of grown or growing horns. Dehorning can be quite painful and even dangerous to the goat, and so upsetting to the surgeon that even many trained veterinarians won't do it, and those who do it once often won't repeat the performance. It's not a job for the casual or beginning goat raiser, either.

Disbudding is quick, easy, and relatively painless if done before the kid's nervous system is fully developed, although it might not appear so to the neophyte. Some European countries, such as Great Britain, require a local anesthetic before disbudding, but the multiple needle insertions and subsequent injections are often more painful and frightening to the kid than the 15 seconds it takes to "burn" each horn bud. The best time to disbud a goat is as soon as you can distinguish the little bump where a horn will grow, usually when the kid is a few days old.

An electric disbudding iron is used to cauterize the horn buds of young kids, so the horns never grow. This is much easier — on you and the goat — than removing horns after they have developed.

The Disbudding Iron

The recommended tool is the electric disbudding iron. Kid-size disbudding irons are available from goat supply houses.

How to Disbud with an Iron

1. **Get the iron hot enough to "brand" wood with little pressure.** Wear sturdy leather gloves.

2. **Hold the kid securely on your lap.** (If you'll be doing a lot of disbudding, eventually you'll want to construct a kid-holding box, which is also handy for tattooing and other tasks. See page 117.)

If you have more than a few kids to disbud, or don't have a helper who can hold them, a disbudding box to restrain the goat can be helpful.

electric disbudding iron

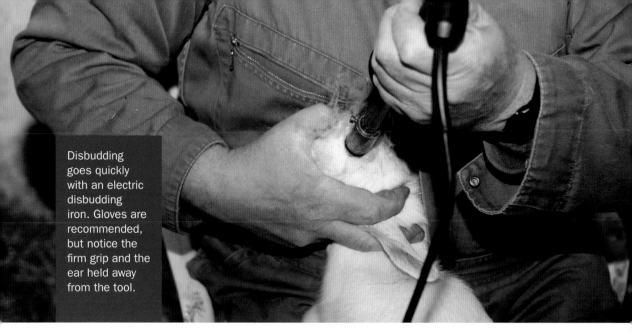

Disbudding goes quickly with an electric disbudding iron. Gloves are recommended, but notice the firm grip and the ear held away from the tool.

3. If the horn has not yet erupted or you're not too sure of yourself, **trim the hair around the horn button with small scissors.** Some people do this routinely.

4. Holding the kid firmly by the muzzle with its ears tucked out of the way, **press the hot iron over the button, and hold it there to a count of 15 and no longer.** There will be acrid smoke from burning hair, a few seconds of violent struggling (which isn't too violent with a kid weighing 10 to 12 pounds [4.5 to 5.5 kg]), and maybe some screaming. But when the 15 seconds are up, everything will be back to normal, except maybe your heartbeat.

5. **Console the kid, and compose yourself** while the iron heats up again, then do the other horn button. When correctly done, the cauterized ring should be a continuous copper-colored circle.

6. After it's all over, **offer the kid a bottle of warm milk or send her back to her mother.** It will take you much longer to forget the experience than it will the goat, but it will be much easier next time.

I knew one hardy, homesteader-type lady who didn't have electricity and who heated a metal rod in her woodburning stove for disbudding.

Proper burn pattern for a buck kid

Caustic

Another method, less hair-raising but also less successful and potentially more dangerous, is to burn the horn buds with a caustic dehorning paste. Several types and brands are available from farm-supply stores and mail-order houses. The directions on caustic are written for calves; ignore the details about when to do the job.

How to Disbud with Caustic

1. **Clip the hair around the horn buttons,** as shown by the dotted lines in the illustration below.

2. **Cut disks of adhesive tape** to temporarily cover the buttons.

3. **Apply petroleum jelly** around the buttons to protect the skin from the caustic.

4. **Remove the tape, and apply the caustic** to the horn buttons.

5. **Secure the kid in an isolated area** for half an hour so other kids won't lick the caustic and so the treated kid doesn't rub caustic on the other kids, its mother, or other parts of its own body.

Clip hair around horn buttons

Caustic can cause blindness if it gets in the eyes, and it will be quite painful on other parts of the body, including yours. One lady I know holds the kid on her lap while watching television for the half hour.

Caustic might seem easier, or less traumatic, than the hot iron — mostly for the person performing the procedure. And the hot iron might seem to be cruel and unusual punishment for the kid. In reality, the iron is more humane. The first second will desensitize any active nerves, and it is all over in half a minute rather than half an hour of acid slowly eating into the skin. Allowing the horns to grow might well be the cruelest alternative,

BREEDING FOR HORNLESSNESS

Some goats are naturally hornless, or polled. So many people ask, why not breed goats for hornlessness?

One of the main reasons has been the genetic link between hornlessness and hermaphroditism. Many goats born of hornless-to-hornless matings are hermaphrodites, or of both sexes, which from a practical standpoint means they are sexless. Horned or disbudded goats can produce polled offspring, but because disbudding generally takes place before the horns erupt, you can't always tell if your kid is polled. A horned kid generally has swirls of hair where the horn would grow if not removed. A polled kid has a decidedly ridgeless forehead and smooth hair that looks like a little cap that comes down to the eyebrows. If in doubt, go ahead and disbud where you think the horn buds are.

if those horns someday tear a gash in another animal or put out an eye.

Disbudding Bucks

The horn buds on a buckling are shaped like teardrops, with their tips pointing roughly to a spot between the animal's eyes. If you use a round disbudding tool, you're likely to miss part of the future horn, resulting in scurs (see below). The trick is to burn each bud twice in an overlapping diagonal figure eight. Be sure the iron is hot enough for each consecutive burn and that you end up with copper-colored circles.

Scurs

Bucks have more stubborn horn buds than do does, and there is also a difference in breeds. If scurs, or thin, misshapen horns, start to develop after disbudding, merely heat up the iron and do the job over again. In some ways, scurs are more dangerous and troublesome than horns are. It's not unusual for them to curve around grotesquely and grow into an animal's head or eye, and thin ones will be broken off repeatedly, resulting in pain and loss of blood. Unfortunately, scurs don't show up until the kid is much bigger and harder to handle, so it's better to be thorough and not take shortcuts with disbudding.

Dehorning

In some cases it might be necessary to dehorn a goat. For example, if you have a herd of hornless goats and bring in a new animal with horns, she's sure to cause problems.

Grown horns can be sawed off, usually with a special flexible-wire blade. Each horn must be removed close to the skull, and a thin slice of the skull taken with it, or the horn will

> **TIP**
>
> Always keep blood-stop powder in your barn medicine cabinet. If you don't have any, use a handful of cobwebs or a liberal sprinkle of cayenne pepper (capsicum or red pepper) in an emergency. Keep pepper and powder out of the goat's eyes, as they can burn.

grow back. There will be a great deal of blood, the decidedly unpleasant view into the sinus cavities of the goat, the real risk of having to deal with infection, and the obvious difficulty of controlling an adult animal. An anesthetic will be required. This is a job for a veterinarian, and as mentioned, most of them don't want to tackle it.

Again, there is an alternative, which, while it might sound easier and more humane, actually isn't.

If a couple of very strong rubber bands — the elastrator bands used for castration (see page 188) — are placed tightly around the base of the horn, the horn will atrophy and fall off. Some people file a notch in the horn very close to the skull to keep the bands down where they belong, and others claim putting duct tape over the bands will hold them on. In any case, check the rubber bands regularly to make sure they haven't broken or moved.

The problems arise when the horn structure begins to weaken. A goat may butt another or merely get the horn caught in a manger or other obstacle and break it off. If it's not truly ready to fall off, there will be considerable pain and a great deal of bleeding. Stop the bleeding immediately with blood-stop powder. If the horn is dangling, steel yourself and cut or pull it the rest of the way off. A brief zing of pain is much better than continual zings every time the goat bumps its flopping horn into something. With proper management — perhaps isolation of the animal so treated, the removal of all obstructions, and very frequent inspection — the rubber-band method seems preferable over sawing, in most cases. But disbudding the young kid is far better, for all concerned.

Tattooing

Tattoos are permanent identification numbers that can help your record keeping, can identify a goat long after you sell it, and are often the only way the police have of identifying a lost or stolen animal so it can be returned to the rightful owner. If you raise registered animals, you'll have to tattoo them. If your goats aren't registered, tattooing is still a good idea.

Tattoos can record a great deal of information, including the animal's age, your farm identification, and a unique number for each goat in the herd. Breed organizations have very specific herd identifiers and ways they want tattoos applied. If you have a registered animal — say, an Alpine you want registered with the American Dairy Goat Association — contact ADGA for instructions and letters that will identify your herd. The herd ID goes in the goat's right ear (or right tail web on a LaMancha), and the individual animal identification goes in the goat's left ear (or left tail web). Federal and state regulations are changing the way of (and the need for) identifying food animals, so you may be automatically assigned a unique set of letters or numbers to identify your farm. Those are best placed in the right ear.

It is up to you to decide how you want to identify an individual animal within the herd. The American Dairy Goat Association uses a letter of the alphabet to indicate the year the kid was born, and then follows it with a number. For example, the letter for 2016 was H, so kids born that year would have a left ear tattoo of H1, H2, H3, and so on. The letters G, I, O, Q, and U aren't used to avoid confusion with other letters and numbers, so 2017 kids would have J, 2018 would have K, and so on.

Tattoo sets are available from farm-supply stores, mail-order houses, and small-animal equipment dealers.

Get ¼- or ⁵/₁₆-inch (about 0.5 or 0.75 cm) tongs and letters. You don't need to buy the whole alphabet and all numbers at once; purchase only those needed to identify your farm, enough alphabet letters for the next 3 or 4 years, and digits to cover the number of kids you think you'll have. Use green ink, as it shows up even on dark-colored animals. By the way, if you have trouble reading a tattoo, try holding a flashlight behind the ear in a darkened building.

tattoo set

Tattoo goats soon after birth. An experienced goat owner can tattoo a kid without help, but an extra set of hands or a kid-holding box will make the job go faster. If it's necessary to tattoo an older animal, fasten it in a stanchion or milking stand.

Note: Plastic ear tags can also be used for identification. They are very popular in large meat herds and the occasional milking herd whose owner can't spend time catching animals to look at tattoos. Although ear tags may be practical in those applications, they do carry risk: goats' ears are thin and tender, making it easy for the tags to rip out.

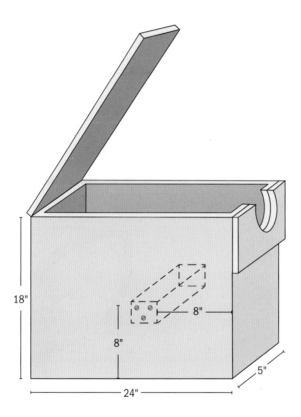

This kid-holding box, made of ¾-inch plywood, keeps the kid still for disbudding and tattooing. Nail or screw a 2" by 2" block of wood across the inside width of the box to keep the kid from lying down or backing up. It can be padded with a towel, if you wish.

18"

8"

8"

5"

24"

YOU PUT IT WHERE?

"Tattoo in the ears?" people asked Judy Kapture, *Countryside* magazine's former goat editor. "Doesn't that hurt?" Judy just waggled her dangling pierced earrings at them and smiled!

How to Tattoo

Insert the letters or numbers in the tongs, remembering that they will read backwards, like a mirror image. Test them on a piece of paper before using them on the goat. (One goat owner I know does the practice tattoo on a file card that becomes the first page of the goat's herd health record.)

1. **Clean the area of ear or tail web to be tattooed** with a piece of cotton dipped in alcohol.

2. **Smear a generous quantity of tattoo ink over the area.** Paste ink can be applied from the tube; liquid ink usually has a roller ball dispenser.

3. **Place the tattoo tongs parallel to the bottom edge of the ear, and puncture the skin with a firm, quick squeeze.** Stay away from warts, freckles, and veins. On very thin-eared kids the needles might go all the way through the ear. Gently release the ear, and pull the skin free. Some tattoo tools have an ear-release feature that eliminates this problem.

4. **Put some more ink on the tattoo area, and rub it into the piercings thoroughly with an old toothbrush or gloved finger.** It takes about a month for the tattoo to heal thoroughly. Infections from tattoos are so rare as to be nonexistent, but tattoo letters should be cleaned thoroughly with a small brush and a little alcohol before storing.

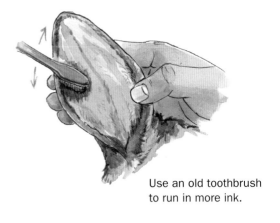

Use an old toothbrush to run in more ink.

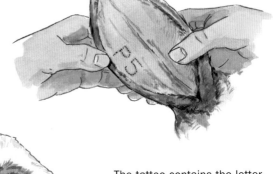

The tattoo contains the letter-designated year of birth and additional information, such as the herd identification number assigned by a registry. Keep a record for yourself.

Place the tongs parallel to the bottom of the ear, and use a firm grip and a quick squeeze.

Hair Care

People who work around food wear hairnets. Since that's not practical for goats, trimming and brushing are important to keep dirt, bedding, and loose hairs from falling into the milk bucket.

Very shaggy goats should have their hair clipped, especially around the flanks and udder. Even a short-haired animal should get a dairy clip, which is basically a tight trim of any part of the goat's body that is above the bucket — the belly, insides of hind legs, and the entire udder right up to the base of the tail. It's a good idea to confine trimming to the udder region in the winter, to keep your goat warmer, but even that "just a little off the top" (or bottom, in this case) will keep your milk cleaner. The entire animal can be clipped in the spring to keep it cleaner and cooler and to discourage parasites. Just like a haircut, a nice clipping will greatly enhance the appearance of a show goat.

Electric livestock clippers are perfect for the job, but they come with a price tag. Electric clippers used for human haircuts work well on the udder, where you want to be right down to skin, but use an attachment that will leave at least $^3/_8$ of an inch (about 1 cm) of hair if clipping the body. These clippers often can be found at rummage sales and thrift stores. Manual dog clippers will work, and if you only have one or two goats, there's no reason you can't use scissors. Just be very careful.

WHAT ABOUT COLLARS?

A collar on a goat is a handy tool for catching and leading, but it can also be a danger, especially if your animals spend time on pasture where the collar can get caught in branches or on a fence. Garden supply stores and goat supply catalogs are good sources for plastic break-away connectors and chain that can be cut to length. Use lightweight chain for kids and the heaviest weight for bucks. In commercial herds, plastic numbers are hung from the chains to identify goats from a distance.

8

HEALTH

Goats are among the healthiest and hardiest of domestic animals. If you pay close attention to proper feeding and other management details, you're likely to have very few health problems with your goats. This chapter gives you a brief A-to-Z introduction to some of the health problems you might hear about or encounter.

The Natural State

It seems to me that some people are obsessed by disease and sickness. They waste money, worry unnecessarily, and don't enjoy life with goats as they should. My own views are different, and I'll explain my attitude so you'll be aware of my bias. Then if you don't agree, perhaps you can find a more technical manual, written by a veterinarian, or at least by someone who shares your interest in sickness (*Goat Medicine*, 2nd ed., by Mary C. Smith and David M. Sherman, is highly recommended).

In my view, sickness is only an absence of health, and health is the natural state. If your animals get sick, it's because of wrong conditions of feed, environment, or in some cases breeding. Treating the symptoms will help in the short run, sometimes, but unless the underlying causes are corrected, any time and money spent on medication is wasted.

What's worse, many illnesses have purposes, and by "curing" them we sometimes compound the problem. Scours or diarrhea is one example. It's fairly common in kids and can result from feeding too much milk or cold milk (when the kid isn't used to it) or using dirty utensils. You don't want to stop the diarrhea cold because that's nature's way of getting rid of the toxins. So you let it take its course while removing the cause: the excess milk, the cold milk, or the unclean utensils. (There are caveats to this example, but for those, keep reading.)

Similarly, a completely worm-free goat is a near impossibility and not a desirable goal under any circumstances. The number and amount of vermifuges (dewormers) required would do more harm than good, and some internal parasites are symbiotic — the goat needs them to live.

We have been led to believe that all microorganisms are bad per se. Nothing could be further from the truth. Even most pathogenic organisms will have little or no effect on a healthy body; only when the host is weakened because of some other factor, such as poor nutrition, does the pathogen get out of hand. Some bacteria are apparently harmless, and some are actually necessary.

Goats tend to be healthy animals, but a small supply of everyday emergency items is good to have on hand. Don't forget an adhesive bandage or two for yourself.

Your job, then, is to maintain the natural state of your goat's health by providing her with the proper feed and environment.

Finding and Using a Veterinarian

This isn't meant to imply that goats never get sick, that if they do it's because we did something wrong, or that there's nothing we can do for them. While you might go for years without seeing any health problems, if you have a large number of goats (or live with a few of them long enough), you're almost certain to encounter some ailments. However, you don't need a medical degree to raise goats. If an animal gets sick, all you need is the phone number of a veterinarian.

One of the most common complaints I hear from goat people is, "My vet doesn't know anything about goats." It's true that a few vets just don't care about goats. Others simply lack experience with them because of their low population and generally good health. In any event, anyone who has graduated from veterinary school knows more about animal diseases than the rest of us do. I believe in making use of their knowledge. And if you have a good working relationship with a capable veterinarian, he or she will be glad to share much of that knowledge with you.

Do try to find a large-animal veterinarian, one who treats sheep and cattle, not cats and dogs or pet birds. And it's a good idea to start a professional relationship by engaging a veterinarian for help or advice concerning routine health maintenance before you desperately need help at 2 o'clock on a stormy morning.

Smith and Sherman, in their second edition of *Goat Medicine,* caution veterinarians that "hobby farmers often perceive goats more as companion animals than as livestock production units. While they may seek the expertise of a livestock clinician, they often expect the 'bedside manner' of the companion animal practitioner." In other words, the veterinarian

A VALID VCPR

Few medical products, such as antibiotics or dewormers, are officially approved for goats, because of the expense of testing and the small market potential for manufacturers. In addition, antibiotic resistance is showing up in both human and animal populations. It is that antibiotic resistance that has pushed the U.S. Food and Drug Administration to roll out new rules to protect the world's available arsenal of medically important antimicrobials (antibiotics). Among them is the requirement that a veterinarian provide written orders for the use of any medically important antimicrobial given to livestock in their feed or water. This includes some of the most common treatments for scours and respiratory ailments in goats. California has already taken steps to remove injectable forms of medically important drugs from over-the-counter availability, and the FDA is expected to follow suit nationally.

To use any of the restricted medicines or to use medicines not specifically formulated for goats, owners need to develop what is called a valid veterinarian-client-patient relationship (VCPR). Don't wait until you are in dire straits to find a veterinarian who is willing to work with you. It might even provide an opportunity for your vet to brush up on goat skills.

has no way of knowing whether you are caring but pragmatic or if you have raised your goat to a status equal to your firstborn. Neither emotion is right or wrong, but be fair, and let the veterinarian know where you stand. Your goat will still get the best care possible, but you will probably have a better working relationship at the next visit.

How to Tell If Your Goat Is Sick

When you've had your goats for a while, you'll get a feel for what they look like when they are healthy. The most obvious way to tell if one is sick is if she doesn't look "normal." She will be off her feed, standing in the corner away from the action, not chewing her cud, shivering or standing hunched up, and looking miserable. More obvious signs are green mucus from the nose, discharge from the eyes, and droopy ears, although that last one is tough to tell with Nubians. Certain parasites and kidding problems have their own sets of symptoms that either will be obvious or will be discussed under their individual headings.

It's a good idea to introduce your veterinarian to your herd before you need an emergency visit. Sometimes a wellness visit can head off later serious problems.

BASIC PHYSIOLOGICAL DATA

Parameter	Normal
Temperature	101.5–104.0°F (38.6–40°C) (rectal). Varies with air temperature, exercise, excitement, and amount of hair. To determine abnormal temperature, compare with several others under the same conditions.
Pulse	70–80 per minute
Respiration	12–20 per minute
Puberty	Occurs at 4 to 12 months
Estrous (heat) cycle	21 days (average); range, 18–23 days
Length of heat period	18–24 hours (average); range, 12–36 hours
Gestation	148–153 days; average, 150 days
Birth weight	8 pounds (3.5 kg)

An A-to-Z Guide to Common Health Problems

Here's a brief overview of some of the conditions you're most likely to encounter, with tips on treatment and prevention.

Abortion

Abortion is the premature expulsion of a fetus and can be caused by a variety of mechanical and medical reasons.

If abortions occur early in pregnancy, the cause is apt to be liver flukes or coccidia. Liver flukes are a problem primarily in isolated areas such as the Northwest, where wet conditions favor them. Coccidia can be transferred by chickens and rabbits, both of which should be kept away from goat feed and mangers (see Coccidiosis, page 128).

Abortion is more common in late pregnancy. The cause can be mechanical, such as the pregnant doe's being butted by another or running into an obstruction such as a manger or a narrow doorway. It can also be related to moldy feed, high nitrate concentrates in the water, heavy feeding of pine trees, or a few unfortunate diseases that occasionally pop up, like Q fever, chlamydia, or toxoplasmosis. Those tend to cause abortions in several animals at once, at which point you would be wise to call on the help of a veterinarian.

Certain types of medication can cause abortion, including worm medicines and hormones, such as those contained in some antibiotics. Medicate pregnant animals with caution.

Abscesses

An abscess is a lump or boil, often in the neck or shoulder region, that grows until it bursts and exudes a thick pus. There are several types of abscesses with different causative organisms and very different degrees of seriousness. Most are related to wounds, including punctures by thorny vegetation, bites, and even hypodermic needles. An abscess can occur when a goat bites her cheek. But the form getting the most attention by far nowadays is caseous lymphadenitis, or CL. It's caused by *Corynebacterium pseudotuberculosis*, formerly known as *Corynebacterium ovis*.

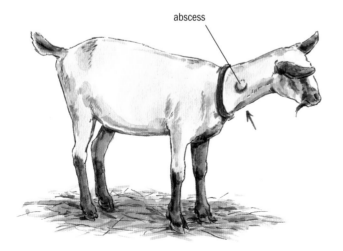

abscess

This doeling has an abscess on her neck. Although this condition is quite common in some herds, it can be considered a serious problem and requires special attention.

The condition can be transmitted from one animal to another, so abscesses are common in some herds and nonexistent in others. In commercial herds, abscesses are often identified as the number one health concern.

First isolated in sheep in Australia in 1894 (hence the name *C. ovis*), CL was seldom mentioned in the United States until the 1970s, and it didn't reach England until the 1980s, with the importation of Boer goats. Today, many consider CL to be the major disease problem of dairy goats in the United States. The late Dr. Samuel Guss, a well-known and highly respected authority on goat health, called a goat with a discharging abscess "a hazard to other goats and to humans." One problem with CL in particular is that abscesses may be forming and draining internally but can't be identified until the animal is dead and has undergone a necropsy.

Treatment

Any animal with an abscess should be isolated. The milk from such an animal should be pasteurized, according to Dr. Guss, and if the abscess is on the udder, the milk should be discarded. Don't feed it to kids; dump it.

The lump can become the size of a walnut, a tennis ball, or even larger. The skin over the abscess will gradually thin until it bursts open, thereby spreading the infection. To avoid this, the abscess should be properly lanced and cleaned before it ruptures. This is a matter of timing. It should be lanced when the hair starts to fall off the skin over the abscess. Wait too long, and it may break open on its own; rush in too early and the abscess may not be differentiated enough for you to do a thorough job.

Clip the hair around the abscess, and disinfect the surface with an antiseptic such as tincture of iodine or povidone-iodine. With a sharp, sterilized knife or surgical scalpel, make a vertical incision as low as possible on the abscess to promote drainage. Wearing disposable latex gloves, squeeze out the pus and burn the material and all cloths that come into contact with it. Flush the wound with diluted disinfectant. Isolate the animal until the lesion is healed and covered by healthy skin, typically in 20 to 30 days.

If the abscess is in the throat, behind the jaw, or under the ear, a veterinarian should perform the procedure, since these areas contain major blood vessels and nerves. But there are other reasons to engage expert help. Most abscesses are zoonotic, meaning the bacteria can be transmitted to humans.

If the abscess is caused by lymphadenitis, it will have a greenish cheeselike pus. A yellowish pus of mayonnaise consistency suggests *C. pyogenes*, while *Streptococcus* often produces a watery discharge and *Staphylococcus* causes a creamy exudate. However, these are only clues. The only way to know for certain is with a lab culture.

Prevention

There has been some work and, obviously, a great deal of interest in vaccinations, particularly ones developed in other countries but not released in the United States. One U.S. company developed a CL vaccine for sheep, but it is an off-label use for goats and has reportedly caused severe, although temporary, reactions in goats. If you had several hundred goats and a serious CL problem, you could have an autogenous vaccine created from pus samples out of your own herd. It is probably not cost-effective for a small herd and still leaves you with infected animals. Control consists mainly in treating and isolating (or culling) infected animals. Remove kids from affected does at birth and raise them on colostrum and milk from clean does.

Herds that are free of abscesses generally stay that way until a new animal is brought in or the goats come into contact with others in some other way. But once established in a herd, CL is difficult to eradicate. If you are taking your doe to a buck or borrowing him for breeding, give him a quick once-over to be sure he doesn't have a draining or ripe abscess that might pop while he's working.

Bang's Disease (Brucellosis)

Brucellosis is a contagious disease primarily affecting cattle, bison, elk, swine, sheep, goats, and dogs, and it is characterized chiefly by abortion. The first *Brucella* infection to be recognized was caprine brucellosis, or Malta fever, in 1887. While it's prevalent in many countries where goats are common, caprine brucellosis is extremely rare in the United States. The last outbreak was reported in Texas in 1999.

Abortion occurs in about the fourth month of pregnancy. Diagnosis requires a bacteriologic examination of the milk or the aborted fetus or a serum agglutination test. Experimental work with sheep and goats found success with an intensive treatment with an antibacterial cocktail, but it is experimental and would only be sensible to attempt with very valuable animals. Even as rare as brucellosis is in the United States, goats are regularly tested.

Many people are concerned about brucellosis, but many years ago *Countryside* magazine investigated the thirteen cases reported in U.S. Department of Agriculture (USDA) annual statistics. These cases occurred in three herds, in Arizona, Indiana, and Ohio. A check with officials in Arizona and Indiana showed that their cases were in fact clerical errors. The one goat in Ohio was classed as positive on the test, suspicious on the retest, and after being slaughtered and subjected to a tissue test, found to be negative.

Goats that have come up suspect after a Bang's test have invariably been pregnant or recently freshened. Subsequent tests are negative.

Nevertheless, to be absolutely certain about the safety of the milk from your goats, you can have them tested for Bang's, and you can pasteurize the milk.

Bloat

Bloat is an excessive accumulation of gas in the rumen and reticulum resulting in distension. Bloat is caused by gas trapped in numerous tiny bubbles, making it impossible to burp. If you've just turned the goat out on a lush spring pasture or if she figured out how to unlock the door to the feed room, anticipate bloat.

A more common problem is bloat in young kids, especially when they are being fed milk replacer rather than goat milk (see chapter 12 on raising kids).

Treatment

A cup of oil given internally — corn, peanut, or mineral — will usually relieve the condition. A handful of bicarbonate of soda also will help. In life-or-death cases, it may be necessary to relieve the gas by making an incision at the peak of the distended flank (on the goat's left side), midway between the last rib and the point of the hip and holding the wound open with a tube or straw.

ALTERNATE BLOAT REMEDY

Give a mineral-oil enema, followed by lots of body-temperature water.

Prevention

As always, the best cure is prevention. Feed dry hay before letting animals fill up on high-moisture grasses and clovers. Don't feed great quantities of succulents such as green cornstalks if the animals aren't used to them. Use a high-quality milk replacer for kids, and let the bubbles go flat before feeding it. If you are feeding a kid in your lap, hold the bottle far enough away so the kid has to stretch its head out and upward so swallows of milk will take the route to the omasum and not to the rumen, where it doesn't belong.

Brucellosis

See Bang's Disease on page 126.

Caprine Arthritis Encephalitis

In the early 1970s, caprine arthritis encephalitis (CAE) was little more than a goat-world rumor. While a few breeders became almost hysterical, most thought it was much ado about nothing.

Research in intervening years has yielded numbers anywhere from 34 to 80 percent of animals infected within the groups tested, but it is nearly impossible to get a good handle on the actual number of animals infected with CAE virus (CAEV) across the country. One of the challenges is that the most common CAE tests — agar gel immunodiffusion (AGID) and enzyme-linked immunosorbent assay (ELISA) — are said to have only an 80 percent accuracy rate, and the most accurate tests are expensive. What is known about the disease is that prevention is the best cure.

Symptoms

In one common form of the disease, the first sign is usually minor swelling in the front knees. The swollen knees become progressively worse, and the animal just seems to "waste away." The lungs may become congested, and eventually, all the body's systems give up. Other symptoms involve chronic progressive pneumonia, a hard udder, and weight loss associated with chronic disease. Less common is an ascending paralysis in kids that otherwise appear healthy.

Treatment and Prevention

Caprine arthritis encephalitis is an incurable contagious disease, and some goats that have the virus do not show symptoms but are still carriers. There is no vaccine. The only recourse at this time is a prevention program.

You can have your goats tested for CAE. If they test positive, there is nothing you can do for them, but because the infection is spread in the neonatal period, you can build and protect a "clean" herd by following this regimen:

1. Be there when the kids are born. Deliver them onto clean bedding, preferably newspaper rather than straw. Don't break the amniotic sac before the kid is delivered: the fluid in the sac is not infected, and the sac prevents the kid from swallowing or inhaling infected cells. Do not let the doe lick the kid.

2. Put each newborn kid into clean, separate boxes so they can't lick one another. As soon as you can, wash each one in warm, running water to eliminate the possibility of any infected fluid on the body being ingested. Keep them separate until they're clean and dry.

3. Within half an hour or so, feed the kids colostrum — from a goat you know is free of the virus or from a cow or, in case of emergency, a commercial or home-brewed colostrum replacer (see page 176). Note that cow colostrum has recently come under fire because of the possibility of introducing other diseases, such as Johne's (see description on page 130).

4. Feed only pasteurized milk or sheep or goat milk replacer (milk replacer made for calves is not high enough in fat for sheep or goats).

5. Keep the kids separated from other goats and practice strict hygiene, including not only sanitized feeding utensils but also such precautions as caring for the kids before handling or walking among the older animals and washing your hands, changing boots, and so on before going among the kids.

Caseous Lymphadenitis

See Abscess, page 124.

Coccidiosis

This disease is caused by microscopic protozoans (coccidia) found in the cells of the intestinal lining and is therefore a parasitic disease (see also Parasites, Internal, page 134). It usually occurs in kids 1 to 4 months old and usually in crowded and unsanitary pens. The most common symptom is diarrhea that turns bloody if left untreated, although kids with coccidiosis are usually weak and unthrifty. Your vet might recommend a sulfonamide in the feed mixture or other coccidiostat in the water to treat this condition. Lower dosages of coccidiostats are often used as a prophylactic. Good management of the kid pen is a more effective and long-term treatment.

SHOULD YOU WORRY?

How concerned should you be about CAE? That depends. The most common tests, ELISA and AGID, are notorious for false negatives that require retesting, but the cost is relatively low, anywhere from $8 to $18 per sample, plus the cost to have your veterinarian draw blood. A newer polymerase chain reaction (PCR) test detects the virus's genetic material (genome) in the white blood cells in a sample. The PCR test for CAEV is significantly more expensive at $39 up to $85 per sample. It may not be cost-effective for most backyard dairy goat owners or even necessary if they raise their own herd replacements or use all kids for meat. On the other hand, breeders who sell goats (particularly across state lines) or who attend shows that require health certificates should use a CAE prevention program as outlined above. There is no evidence that milk produced by CAEV-positive animals can affect humans.

Cuts

Cuts, punctures, gashes, and other wounds can almost always be avoided by good management. They can be caused by such hazards as barbed wire, horned goats, junk, or sloppy housekeeping and other conditions under the control of the goat caretaker. Still, accidents can happen.

Clean such wounds with hydrogen peroxide and treat with a disinfectant, such as iodine. Use your own judgment to decide if stitching is required, or get the animal to a veterinarian. Verify that its tetanus vaccination is current. The tetanus vaccine is often found paired with vaccines for *Clostridium perfringens* types C and D and is commonly called CDT.

Diarrhea

See Scours, page 138.

Enterotoxemia

Enterotoxemia is also called pulpy kidney disease and overeating disease. An autopsy soon after death will often show soft spots on the kidney.

The usual symptom of enterotoxemia is a dead kid, although it is not unheard of in unvaccinated adult goats. There is always misery and almost always a peculiarly evil-smelling diarrhea. With some strains, there may be bloat or staggering, starting with weakness in the hindquarters and quickly spreading to the whole body.

It is caused by a bacterium that is always present but that, when deprived of oxygen in the digestive system, produces poisons. There are six types of *Clostridium perfringens* bacteria that cause enterotoxemia. Types B, C, and D cause the most trouble, with type D most often affecting sheep and goats. The proper conditions in the gut can be induced by overfeeding or making a sudden change to the goat's diet. Goats build up resistance to the poisons produced in small, regular amounts, but they can't handle sudden surges of them.

Treatment

Antitoxin can be administered if you get there fast enough, but death is usually swift. Where enterotoxemia is a problem, vaccines are available from your veterinarian (see Immunizations in appendix, page 261). They are cheap, easy to administer, and much easier to deal with than dead goats.

Prevention

The best prevention is proper feeding on a suitably bulky, fibrous diet. If you must change your grain mix, do it gradually over a week or so. Annual CDT booster shots should be given to the doe 45 to 60 days before kidding. That will give her kids a slug of antibodies that will keep them relatively safe until they are ready for their first vaccination at about 6 weeks of age.

Studies over the past decade have found that most commercially available vaccines — designed for sheep but used on an extra-label basis for goats — did not maintain the proper levels of protection that they did for sheep. If you are in an area where enterotoxemia has been a problem, your veterinarian may recommend that animals be vaccinated at least twice a year, and a vaccination schedule of every 3 to 4 months would give even better protection.

Floppy Kid Syndrome

This is a relatively new term for goat owners and refers to a kid that, at 3 to 10 days of age, loses its appetite and seems to wobble and have no muscle tone when you pick it up. It is caused by clostridial-type bacteria that favor warm weather and an excess of milk.

Bottle-fed kids have distended stomachs and "slosh" when you move them. Nursed kids are hollow, and their mothers are overfull, as though they haven't been nursed. The kid often perks up after a few days of receiving doses of sodium bicarbonate or pink bismuth, but flat, unresponsive kids may need electrolyte therapy and veterinary care. Never try to force milk into an unresponsive stomach.

Goat Pox

The symptom is pimples that turn to watery blisters and then light-colored pustules and encrusted scabs on the udder or other hairless areas such as the lips. It varies in severity.

Pox can be controlled by proper management, especially that involving sanitation. Infected milkers should be isolated and milked last to avoid spreading the malady to others. Time and gentle milking are the best cures. Traditional treatment is methyl violet to dry up the blisters, but this can make the udder very dry and painful.

Prevention

Very similar symptoms can be caused by irritation. I have seen cases caused by dirty, urine-soaked bedding and by the use of udder-washing solutions that were too strong. In all cases the cure is wrought by removing the cause. An antibiotic salve will keep the skin supple and prevent secondary infections.

Hoof Rot

This disease shows up mostly in warm, wet southern climates but is not unheard of during the summer in the north. It is most often associated with overgrown hooves and continuously wet conditions, although bacteria can also enter the hoof from injury or contaminated discharge from another infected animal.

Prevention

Keep barnyard and pens clean and dry. Where goats have access to swampy land or continually wet pasture, keep hooves trimmed so manure and mud cannot become impacted. If hoof rot is a problem, consider setting up a footbath of 10 percent copper sulfate in solution. Goats don't like walking through water, so the footbath might have to be a daily hands-on ritual or placed where the goat has no choice but to walk through it. A topical disinfectant or antibacterial can be used on an infected hoof after trimming.

Johne's Disease (Paratuberculosis)

Like CAE, this is one of the "wasting" diseases. It is a bacterial disease primarily affecting the digestive tract, probably with fecal-oral transmission. Frequently, the only symptom is extreme weight loss or gradual loss of condition. Scouring is a typical symptom in cattle but uncommon in goats. Johne's disease (pronounced YO-nees) is hard to diagnose accurately in goats except by autopsy.

The disease apparently infects young animals through manure-contaminated teats or directly from the milk of a heavily infected doe. The infected kid typically won't start getting sick for 1 to 2 years. Older animals presumably contract it from sick animals shedding the organism in their manure. Adult animals can be sources of infection even if they do not show clinical signs of disease. Land previously used by infected cattle can remain contaminated and infect goats that are brought in several years later.

Testing is done by fecal sample or blood test, but both are often misleading if the disease is subclinical in the animal at the time of testing. There are no vaccines available in the United States.

The question of whether Johne's disease can be transmitted to humans is highly controversial. There is no data to show that such a transmission is possible, but research is being done, because Crohn's disease in humans is a very similar chronic inflammatory bowel disease. *Mycobacterium paratuberculosis* has been grown from samples of Crohn's disease patients, but there is no indication that it caused the disease.

Prevention

The best prevention consists of starting with a clean herd and keeping it that way. Beyond that, infected individuals should be identified and removed from the herd, and new infections in susceptible kids should be reduced by improved sanitation and modified kid-raising methods, including isolation from adults. Do not let the doe lick the newborn, and don't allow the kid to nurse.

Ketosis

Ketosis is also referred to as pregnancy disease, acetonemia, twin-lambing disease, and other names. Symptoms include a lack of appetite and listlessness. Urine has a distinctive nail-polish smell. Ketosis occurs during the last month of pregnancy or within a month after kidding. Its primary cause is poor nutrition in late pregnancy, but it's most likely to affect fat does, especially those that get little exercise. A dairy goat should never be fat;

nutrition is particularly important when the unborn kids are growing rapidly and making huge demands on the doe.

Treatment

Treatment consists of administering 6 to 8 ounces (175 to 235 cc) of propylene glycol. This may be given orally twice a day but not for more than 2 days. In an emergency, try a tablespoon of bicarbonate of soda in 4 ounces (118 cc) of water, followed immediately by 1 cup (235 cc) of honey or molasses. Once an animal is at the point of having little appetite and then quits eating altogether, there is no effective treatment.

Lice

Suspect lice if your goat is abnormally fidgety, has wet whorls of hair from chewing on her coat, is losing weight for no apparent reason, or has a generally scruffy and dull coat. Regular spring clippings and lots of sunshine are good preventives. For an old-time cure, apply two parts lard and one part kerosene to the coat.

Lice are almost universal, but mild infestations cause little harm to well-nourished animals. They tend to pop up during the winter when the doe is stressed with kidding. A badly infested goat will rub against posts and other objects, will have dry skin and dandruff, and can lose a great deal of hair. Badly infested kids can actually die from stress and malnutrition.

Lice can't tolerate sunlight, so a good spring clipping and a day or two in the sun often takes care of any summer infestations.

When it isn't practical to clip, lice can be controlled by dusting, spraying, or, in large herds, dipping. Effective louse powders are available at farm supply stores, or ask your veterinarian for a louse powder approved for use on dairy animals. All members of a herd must be treated at the same time to control lice, and the treatment must be reapplied within a few weeks to catch the newly hatched nits.

Mange

Mange is indicated by flaky "dandruff " on the skin. It's accompanied by irritation. Hairless patches develop, and the skin becomes thick, hard, and corrugated. The condition is caused by a very tiny mite. There are several types of mange. Demodectic is probably most common and can be stubborn. Mange can be treated with a variety of medications, including amitrol and lindane, available from veterinarians. Follow the directions on the label.

Flaky skin can also be the result of malnutrition or internal parasite infestation.

Mastitis

Mastitis is inflammation of the mammary gland, usually caused by an infection. Symptoms are a hot, hard, tender udder and milk that may be stringy or bloody (routine use of a strip cup before milking will alert you to abnormal milk; see chapter 13 for details). Mastitis may be subclinical, acute, or chronic. It's usually a treatable minor problem, but some forms, such as gangrenous, can be deadly.

Not all udder problems indicate mastitis. Hard udders (typically just after kidding) that test negative for mastitis are just congested and usually soften in a few days. Congested udders are best cleared up by letting the kids nurse and *gently* massaging the teats and udder for 3 to 4 days after parturition. Some goat owners swear by an apple cider vinegar massage as the best cure.

In mastitis, the alveola, or milk ducts, are actually destroyed. Since it's necessary to identify the bacteria involved to know how to treat it, the services of a vet are required.

Mastitis can be caused by injury to the udder, poor milking practices, or transference of bacteria from one animal to another by the person milking. Owners milking by hand should wash their hands between each animal, and teats should be predipped and dried before milking. Teat dips have proved of great value in controlling the disease among cattle, but make sure the one you choose for your goat does not irritate her skin.

Somatic cell counts, used commercially to detect mastitis in cows, are not considered reliable with goats (see appendix, page 257).

Home tests for mastitis are available from veterinary supply houses. The best known is the California mastitis test, or CMT, in which a reagent is mixed with equal parts of milk and

USE A TEAT DIP AFTER EACH MILKING

Goat veterinarian Dr. Joan Bowen says that consistent use of an effective teat dip after every milking could prevent half of all cases of mastitis. She explains that during milking the small sphincter muscle around the teat orifice relaxes, allowing the orifice to open and the milk to drain out. It takes about 30 minutes for this orifice to close again. This is when bacteria can enter the teat canal, causing mastitis. A teat dip kills the bacteria before the orifice closes.

MASTITIS AND LENGTH OF MILKING INTERVALS

You might be interested to know that the widespread belief that mastitis can result when animals are not milked at 12-hour intervals has not been proven. One study has shown that goats milked at intervals of 16 and 8 hours produced as much milk as those milked at 12-hour intervals and with no increase in the incidence of mastitis. Milking should be done at regular times, but it doesn't have to be a 12-hour schedule.

observed for signs of coagulation. The mixture is scored from 0 to 3. No coagulation — and presumably no mastitis — is scored as 0; a thick gel, and presumed mastitis, is scored as 3. Unfortunately, other factors can cause the reagent to thicken, and a positive test may be misleading. A veterinarian can verify a presumptive positive CMT.

Prevention

As you might expect, an ounce of prevention is worth a pound of cure. Dry, clean bedding, especially at freshening, is important. Predip and dry udders with individual towels for each animal. If you wash udders with a sanitizer, use a spray bottle rather than a communal bucket. Milk with clean, dry hands. Milk gently, in peaceful surroundings. Avoid vigorous stripping.

Many goat owners swear by dry treating, which is the infusion of a long-acting antibiotic into each udder half when the goat is dried up at the end of her lactation. Studies on cows have proven that the practice prevents subclinical mastitis for that species, but the panel is still out on its effect on goats. There are several problems, not the least of which is that dry-treat tubes come 12 to a box and could well be outdated by the time a small backyard herder gets through the whole box. Another is that the

dry treatments are off-label use, and there have been no definitive studies to suggest how long milk must be withheld after kidding.

Milk Fever (Parturient Paresis or Postparturient Hypocalcemia)

Symptoms of milk fever include anxiety, uncontrolled movements, staggering, collapse, and death. Usually, this occurs within 48 hours of kidding. It's caused by a drastic drop in blood calcium, which is related to the calcium level of feed consumed during the dry period and even to incorrect feeding of young animals. It can be brought on by sudden changes in feed or short periods of fasting when the goat has great demands on its body late in pregnancy or early in lactation. Curing milk fever requires quick action and a veterinarian who will administer calcium borogluconate intravenously, which is typically quite effective.

Orf (Soremouth)

Orf is a viral disease that is transmitted by contact with infected animals. It often shows up first on the mouth as raised black blisters and scabs but can also be found on other hairless body parts. It is a particular problem for kids, because they may have trouble nursing due to the crusty deposits around the mouth. Or the doe may get sores on her teats and refuse to nurse the kid. Once an animal has had orf, it is immune to future episodes.

The virus is highly contagious to humans as well as goats, so wash hands thoroughly after treating or handling a goat with orf. If the scabs are around the nose of an adult goat and do not clear up with treatment, the problem may be a zinc deficiency, which is a feed problem (see chapter 6).

Treatment

There is no treatment for orf itself, but a topical application of gentian violet, tea tree oil, zinc oxide, or WD-40 will help dry the blisters so they heal faster. Some veterinarians are vaccinating with ovine ecthyma vaccine with some success, but it is a live virus and can infect the handler if not treated with respect. Many goat owners who don't show their animals view orf in the same way that parents view chicken pox: hope everybody gets it at once and gets it over with. Unfortunately, once the virus is on the premises, each year's new kid crop is likely to go through a few weeks of the same treatment.

Parasites, External

External parasites are generally much less of a problem with goats than are internal parasites.

Screwworms can be a problem, particularly in the South and Southwest. They survive on living flesh and normally depend on wounds — including those from dehorning and castration — in order to enter an animal. Castrating with the Burdizzo (see page 187) and timing dehorning so the wounds will be healed before fly season will help prevent screwworm infestations.

Dog and cat fleas seldom bother goats and then only in tropical regions. Goats can be affected by mites, which produce the diseases mange and scabies. Sarcoptic mites, responsible for sarcoptic mange, can affect all species of animals; demodectic mange and psoroptic ear mange are specific to goats. Demodectic mange is generally associated with crowded confinement housing, as the mites survive only a few hours away from the goats. Deer ticks have been known to cause Lyme disease in goats, and keds, which look much like ticks but are a fly, cause irritation and damage to the hide.

See also Lice, page 131; Mange, page 132.

Parasites, Internal

"Worms" of various kinds are perhaps the most widespread and serious threat to goats' well-being but only when they're present in large numbers. Many goat raisers worm goats regularly with a favorite anthelmintic, or wormer, but science tells us that is a surefire way to rapidly grow drug-resistant worms.

Incidentally, most people call it *worming*, but feeding and watering refer to *giving* goats feed and water, and we sure don't want to give them worms! *Deworming* is the better term.

The list of internal parasites that infest goats is quite long. It includes bladder worms, brown stomach worms, coccidia, four species of *Cooperia*, hookworms, liver flukes, lungworms, nodular worms, stomach worms, tapeworms, whipworms, barber pole worms, and others. Some are quite common in certain areas and rare elsewhere. The fact is, some goats are naturally resistant to internal parasites, while others can host a certain load of parasites quite handily without undue damage. To prevent parasites from becoming resistant to vermifuges, the recommendation is *not* to deworm on a regular basis and to treat only those goats that are symptomatic.

Not all worms are affected to the same degree by a specific anthelmintic. This means a fecal test is required so your veterinarian (or you, if you have a microscope and an interest in such matters) can determine which parasites are present and therefore which veterinary product or combination of products to use if it becomes necessary.

Deworming Agents

Some of the more common vermifuges include levamisole (Tramisol), thiabendazole (TBZ; Omnizole), cambendazole (Camvet), fenbendazole (Panacur), mebendazole (Telmin), and oxfendazole (Benzelmin). Injectable, paste, and pour-on dewormers like ivermectin have also become popular because they take care of external parasites, but they are not as effective on the internal parasites because it takes longer for the active agent to get to the gut where the worms are. As if matters weren't complicated enough, some cross-resistance to a few of these drugs in the same chemical family has been reported, and most are extra-label applications for goats and require orders from your veterinarian. And of course, it is important to be aware of dewormers that can cause abortions in pregnant does.

This issue of drug resistance and subsequent research has turned parasite management in goats on its ears. Rather than treating every animal in the herd, the recommendation is to treat only the heaviest and most affected carriers. Better yet, cull the animals that are found to need repeated treatment and keep those that show a natural resistance

to the parasites. Some research suggests the offspring of those goats will be better at resisting parasites.

As for treatment, the protocol used in New Zealand, where they know a thing or two about goats, is to group two or more dewormers to kill the most worms at one time. This method is proving effective in reducing drug resistance and in an Australian study has even indicated reversal of resistance. Unfortunately, these pre-mixed deworming cocktails are not approved in the United States, but your veterinarian can recommend an appropriate mixture. One caution: they are not to be mixed in the same container but can be administered separately one after the other.

Methods of Administration. Anthelmintics come in various forms: boluses (large pills), pastes, gels, powders, crumbles, and liquids. Boluses are popular, but many goats refuse to take them, and because they can choke a goat, some people refuse to use them. They can be administered with a balling gun, or try hiding a bolus inside a gob of peanut butter.

Drenching, or administering a medication from a bottle, can also be risky, but it's almost a required goat-keeping skill. A handy drenching gun is fitted with a slightly curved tube that slides easily between the goat's teeth and cheek. A pump mechanism doses the proper, preset amount. It's important to (1) give a little at a time and allow the goat to swallow in between; (2) give it in the left-hand corner of the mouth; and (3) never raise the head—keep the muzzle level. There is a risk of getting the medication in the lungs.

When administering paste wormers, be sure the goat doesn't have anything in its mouth, including a cud. Put the paste in the back corner of the mouth on the left side. If the goat wants to shake her head and fling the paste out of her mouth, hold her muzzle gently and massage her throat until you're sure the dewormer has been swallowed.

Inspecting Your Goats

Pay attention to your goats' mucous membranes, the gums, and particularly the tissue around the eyes. The "whites" of the eyes should not be white but pinkish or red. If they're white, or if the gums are pale pink or gray, this indicates that the goat is anemic. The likely cause is worms.

Dr. Francois "Faffa" Malan of South Africa developed a chart that shows five levels of anemia in goats according to the color of the conjunctival tissue around their eyes. With the

FAMACHA program, only those animals showing serious anemia are dewormed. Poor keepers or goats that require regular deworming are culled under the program. The program was developed for *Haemonchus contortus* or barber pole worm, which is a serious problem in the southern United States, but even northern goat keepers have been trained to use the chart as part of their integrated parasite management programs.

Looking for anemia isn't foolproof, though. *Muellerius*, for example, doesn't cause anemia; it destroys lung tissue. (Also, it is not affected by the most common goat dewormers. If you simply deworm on a regular basis, you might enjoy a false sense of security.) In addition, many parasites will eventually build up resistance to a given anthelmintic. Consult a veterinarian for specific recommendations.

NATURAL DEWORMING

Many goat owners want to use natural ways to deworm their goats. There are many homeopathic products available, and some have proven good results. The U.S. Department of Agriculture's Agricultural Research Service studied ingested copper oxide wire particles as a way to control barber pole worms in sheep and goats and found doses as low as ½ gram reduced nematodes by 60 to 90 percent for at least 4 weeks. The same researchers found that a patented formulation of Chinese bush clover, or *Sericea lespedeza*, would control barber pole worms in both grazed and pelleted forms. Tannin-rich birdsfoot trefoil has also been shown to control worms in sheep and goats. An excellent resource for integrated parasite management ideas can be found at www.wormx.info.

Fecal Exams

The best advice that can be given to beginners, or to anyone who doesn't want to become an expert on worms, is to have laboratory fecal exams performed twice a year and to follow the advice of a veterinarian. A management program can be set up based on the life cycles of the specific parasites present, which animals need treatment, and the anthelmintics chosen.

Pinkeye

Pinkeye refers to many ailments that can affect a goat's eyes. Some are brought on by viruses, some by bacteria, and others by mechanical or environmental causes, such as dusty hay, blowing sand, or even stress and excess exposure to the sun. A first indication is usually a watery or pussy discharge from the animal's eye. Untreated infectious pinkeye can cause the cornea of the eye to ulcerate, making it look milky. If not treated, the goat can become permanently blind and die from starvation or secondary infection.

Topical and systemic antibiotics do a good job of clearing up infections, but they are all extra-label for goats and should be handled by a veterinarian. Be sure to ask the veterinarian if the antibiotic has a withholding time before milk can be safely consumed. Some types of pinkeye can be transmitted to humans, so be careful. Sunlight aggravates pinkeye and slows healing.

Pneumonia

Pneumonia is a broad term referring to a number of lower-respiratory-tract diseases. The etiology (origin) is often unknown without a report from a microbiologist or pathologist. Until that report is in hand, even a veterinarian often must make an educated guess at a specific diagnosis and appropriate antibiotic therapy. This is usually based on experience, but if one antibiotic doesn't work, another may be used. In addition, multiple factors are often present, including worms. What all this means for the goat owner is that the main thing to know about pneumonia is how to prevent it.

Prevention

Ventilation tops the list. Goats needn't be kept in warm housing, but it must be well ventilated and draft-free. In a closed barn, an engineered ventilation system should be considered essential. In cold climates, newborn kids can be removed from the main barn to more suitable housing or, if left with their dams, dressed in wool coats or body socks (often made from discarded sweatshirts or similar clothing). Plenty of colostrum and dry bedding are needed.

Crowding must be avoided. Wet bedding must be avoided. High humidity and condensation must be avoided. Insulation can reduce ceiling condensation but will increase the need for ventilation. Never line a barn with plastic sheeting, which will increase the humidity. Repair leaking waterers that add to humidity.

Dry, clean bedding is essential, but minimize dust. Put your head down where the goat is breathing. If ammonia fumes make your eyes water and your throat feel scorched, imagine what it is like for the goat. Clean the barn and spread new bedding. Isolate new or stressed animals, including those that have traveled to shows.

Poisoning

Symptoms of poisoning include vomiting, frothing, and staggering or convulsions. Because of the nature of a goat's eating habits, poisoning from plants is rare: a goat takes a bite of this and a taste of that and will seldom eat enough of one poisonous plant to cause much damage. Toxic plants are more common in the western and southwestern states than elsewhere. To learn what plants in your particular locale are poisonous, check with your local county Extension agent. Some to watch out for are locoweed, milkweed, wilted wild cherry leaves, and mountain laurel (see chapter 6).

Lead poisoning used to be a possibility when goats chewed on painted wood or were fed weeds gathered along roadsides contaminated with auto exhaust. Both causes are less common today, due to the banning of leaded paint and gas. Still, avoid taking feed of any kind from along roadsides that might have been sprayed for weed control.

Many old farms had their own dump sites tucked in a copse of trees where field rock, empty pesticide containers, dead batteries, and assorted junk were deposited and forgotten. If your property is like that, check especially for leaking batteries and rusted cans that might contain the remains of toxic contents.

Don't feed Christmas trees to goats unless you are sure they are not sprayed with toxic substances. Even fir greens that aren't sprayed can cause abortion in pregnant does if fed at too high a rate. If your neighbor sprays any crop, keep your goats away from any area that might have been contaminated by drifting spray.

Many seeds are treated and poisonous. Every so often, we hear of fertilizers or insecticides or other chemicals that look like feed additives killing off whole herds of cows when someone mistakenly grabs the wrong bag. Be careful.

Antidotes depend on the poison. Call a veterinarian.

Ringworm

Ringworm is caused by a fungal growth rather than a worm. It appears as a ring of hairless skin with a rough, crusty center. Ringworm is zoonotic, meaning that it can be transferred between humans and goats, so use protective gloves and wash hands well after handling a goat with ringworm. One goat owner I know mixes athlete's foot powder with Vaseline and smears it liberally over the affected area. The spot(s) can also be sprayed with a commercial athlete's foot treatment, WD-40, or Blu-Kote, an antifungal available in most livestock-supply stores. Your method of treatment will be more effective if the crusty surface is scraped off a day or two after the first application and a second coat is applied.

Scours (Diarrhea)

Scours in adult goats is a symptom of many diseases but is occasionally caused by natural bacteria in the stomach dying from too much acid production or from extended use of antibiotics. If you cannot hear active rumbling in the rumen, that is the primary suspect. To restore digestive flora, treat the goat to a few tablespoons (or more) of yogurt containing active cultures or a commercially available probiotic powder.

Scours in newborn kids can indicate any of a number of problems, including failure to ingest colostrum soon after birth, lack of sanitation, inadequate nutrition of the doe during gestation, feeding excessive amounts of milk, and feeding low-quality milk replacers. The mortality rate is high, and swift action is required.

HYPOTONIC VS. ISOTONIC SOLUTION

This electrolyte solution is called *hypotonic,* meaning that it contains electrolytes in roughly half the concentration of electrolytes in the blood. This solution is given only by mouth. However, a veterinarian can administer *isotonic* electrolyte solutions, in which the concentration is the same as in the blood, intravenously.

Homemade Electrolyte Solution

When a kid has diarrhea, it becomes dehydrated, its blood pH becomes more acidic, and the resulting depression often causes it to stop eating, resulting in an energy imbalance. The treatment is to restore electrolyte balance. Electrolyte formulations are available from drug companies or as infant-care products, but in an emergency a suitable solution can be mixed from ingredients found in any kitchen. Here is one home remedy of many.

2 teaspoons table salt

1 teaspoon baking soda

8 tablespoons honey, white corn syrup, or crystalline dextrose (never cane sugar!)

1 gallon warm water

Your veterinarian may recommend that neomycin, nitrofurazone, or chloramphenicol be added to formula, or given separately according to the dosage on the package.

1. Add salt, baking soda, and sweetener to water.
2. Mix well.
3. Add antibiotic to formula if prescribed.
4. If the kid is too weak to nurse, administer with a syringe or stomach tube. Give 1 to 2 cups per 10 pounds of body weight per day until the scours clears up. The kid still needs the energy provided by milk, so don't stop its meals, although it may not have capacity or inclination to eat much.

Home Remedy for Scours

This is another home remedy for scours, provided by a veterinarian.

1 cup buttermilk

1 raw egg

1 teaspoon cocoa

¼ teaspoon baking soda

1. Mix ingredients in a blender, or shake well in a jar.
2. Bottle-feed one-fourth of this mixture every 2 to 3 hours. One crushed bolus of neomycin can be added to this remedy, but only with your veterinarian's order.

Soremouth

See Orf on page 133.

Tetanus

Goats with tetanus, also called lockjaw, will hold their heads up in an anxious posture and will be generally tense. They have difficulty swallowing liquids and have muscular spasms. Death occurs within 9 days. Tetanus requires a wound for the *Clostridium tetani* to enter, but it can be something so simple the caretaker doesn't even notice it. Disbudding, tattooing, fighting, castration (especially with elastrator bands), dog bites, and even hoof trimming can set a goat up for tetanus. Horses and mules are often associated with tetanus, but the spores are widespread in soil and animal feces and can survive for many years.

Treatment

Treat all punctures and cuts with iodine or Blu-Kote, and pay special attention to the navels of newborn kids. Routine tetanus vaccination is recommended. Treating tetanus is a job for a veterinarian, but early identification and treatment are important.

Tuberculosis

Two important resources for goat health information agree that goats can be infected by tuberculosis. Smith and Sherman in *Goat Medicine* say that "contrary to the fervent belief of many hobbyists, goats are susceptible to tuberculosis." They refer to studies that have shown goats can both receive and transmit the disease from and to humans. The *Merck Veterinary Manual* agrees but says the disease is rare in goats. When it does occur, the bovine type causes a condition similar to that in cattle, including potential contamination of milk from viable tubercle bacilli. In many states, goats cannot be transported across borders or taken into a show ring without health papers that include a negative TB test.

White-Muscle Disease (Nutritional Muscular Dystrophy)

This is caused by a lack of selenium or vitamin E. This muscular dystrophy most commonly affects healthy, fast-growing kids less than 2 months old, although problems can occur in mature animals. Diagnosis is often difficult even for a veterinary practitioner. Stiff hind legs can indicate white-muscle disease in kids, but it could be tetanus. Sudden death in young kids could be caused by white-muscle disease or by gastrointestinal parasitism. Some cases show symptoms similar to enterotoxemia. Definitive diagnosis requires a postmortem examination of muscle lesions or examination of blood or tissues.

Soils in many parts of the United States are deficient in selenium. Providing this mineral to the doe 2 to 4 weeks before kidding will prevent deficiencies in the doe, which can show up as retained placenta and weak muscle contractions. Treating the doe will also give temporary protection to the kids, which should receive injections at 2 to 4 weeks. Some goat owners who know they have trouble with selenium

---

HOME FIRST-AID KIT FOR GOATS

Even if you have no intention of dealing with a goat ailment yourself, there are a few items that should be close at hand to tide you over until the veterinarian comes. Among them are scissors, Vet Wrap and gauze, blood-stopping powder, Blu-Kote or gentian violet, alcohol or alcohol wipes, pink bismuth, and mineral oil. The kit can be expanded as your experience grows.

If you intend to give any vaccinations, or in the case that your veterinarian might prescribe antibiotics, disposable needles and syringes should be added to the list. One tool in your kit should be a bottle of epinephrine, but it is now highly regulated. If a goat shows signs of anaphylactic shock or allergic reaction to a vaccine or antibiotic, epinephrine administered at a rate of 1 cc subcutaneously per 100 pounds (45 kg) of body weight may be the only thing to save her life. Consult with your veterinarian for options.

deficiencies will vaccinate the kids as a matter of course within the first few days of birth. There is a relationship between selenium and vitamin E. The two are usually administered together as a product called Bo-Se. It is another product intended for veterinary use only, so your veterinarian will recommend the dosages.

Note that in excessive amounts, selenium is a poison. In some areas soil levels are so high that plants grown on them cause toxicity, resulting in paralysis, blindness, and even death.

Worms

See Parasites, Internal, page 134.

Don't Expect to Be a Goat Doctor

To repeat, if you start with healthy goats and give them proper care, chances are good that they'll have few health problems.

Not that goats don't get sick. But rare as health problems might be, there are so many possible diseases and ailments that it doesn't make any sense for a person with a few animals to try to be familiar with them all, even if that were possible.

If you do have a sick goat, don't hesitate to call a veterinarian. It's easy for amateurs to misdiagnose animal illnesses. Furthermore, if you try to get a veterinary education from a couple of books, try all the home remedies you can find, and only call the practitioner when the animal has three feet in the grave, don't expect the doctor to perform miracles. Veterinarians carry drugs, not Bibles.

AVAILABLE DRUGS AND EXTRA-LABEL USE

While most homestead goat raisers are not keen on using drugs, those who don't share that aversion should know that only a handful of drugs are currently licensed for use in goats.

Certain others can be used but only under the supervision of a veterinarian, known as "extra-label use." Conditions are spelled out in the Animal Medicinal Drug Use Clarification Act under the authority of the Food and Drug Administration. One rule is that a veterinarian-client-patient relationship must exist. It is strictly illegal for nonveterinarians to use these drugs.

9

THE BUCK

Goats won't produce milk without kidding, and they won't kid without being bred. That requires the services of a buck, and that entails a whole 'nuther look at goat raising.

You'll need a buck, but that doesn't mean you'll have to buy one. In fact, beginners are usually advised to forget about keeping a buck and find one that can be borrowed for a month or be available to receive company when one of the girls is in heat. Even if you choose not to own a buck, this chapter is still important, because it will help you better understand what you need to be looking for in a loaner.

Whether — or Not — to Keep a Buck

One practical motivation for not keeping a buck is expense. A buck requires the same amount of housing, feed, bedding, and grooming as a doe. Therefore, if you have one doe and one buck, the cost of your milk is double what it would be without the buck. The number of does you need to justify the expense of having your own herd sire depends pretty much on your particular situation. If you live in a remote place where there is absolutely no stud service available within a "reasonable" distance (and this might be a few miles or a few hundred miles, again depending on you), then it might be necessary to have a buck even for just a few does. It's expensive but more economical than the alternatives.

Of course, if you choose to keep a buck for just a few does, you must decide how you want to deal with the problems of inbreeding in subsequent years when it's time to breed the buck's daughters and granddaughters. While some breeders of high-quality animals intentionally line-breed — skipping a generation; for instance, breeding a doe back to her grandfather — it has the potential of backfiring. Line breeding can just as easily multiply faults as it can strengths. Even with very tight line breeding, you need two unrelated bucks and should use them alternately for each generation. If you have several breeds of goats and want to keep them purebred, you would need a buck or two for each breed. Depending on how eclectic your herd is, that can turn into a lot of bucks — in more ways than one.

Bucks are hard on fencing, buildings, and each other. They need to be kept apart from the does, so you can choose who is bred when and so their less than flowery odor isn't absorbed by milk you are going to drink. Extra housing means extra expense and maintenance.

These young Oberhasli bucks display the wide chests, long and level rumps, and squarely set feet and legs that their owner hopes will pass to the next generation of kids.

Buck-flavored milk isn't too pleasant, and neither is the smell of the air around your house and barn during the breeding season. Bucks stink! And some bucks that were not properly raised or trained can be dangerous, especially to people not physically and mentally equipped to handle them. We'll look at both of these in more detail in a moment.

The obvious reason for having a buck is to breed does so they can produce kids and milk. If you have a lot of does or are intending to expand your herd and aren't interested in taking the time to ferry them to the nearest buck every time one comes into heat, it may be sensible to own your own.

Improving the Breed

Here's the rub: If all you want from your goats are milk and meat, you might assume that any "doe freshener" will serve the purpose. Even if your only interest is being a self-sufficient homesteader, that might be shortsighted. While a doe can be expected to produce one or two kids a year, a mature buck can breed as many as a hundred does a year. In your herd, his genetics will pass to all his offspring and have an impact — positive or negative — on many generations to come.

Most emphasis on breed improvement naturally comes from people who are involved in showing their animals, be they rabbits, dogs, cows, or goats, but there is more than ample evidence to suggest that both "commercial" and backyard producers have every bit as much to gain from striving for improvement, and perhaps more. I have found it frustrating to deal with people who place little or no emphasis on breed improvement or who even actively belittle the "fancy-pants" show enthusiasts as if their interests were somehow contradictory.

Nothing could be further from the truth. If you keep the same two does until they are old and decrepit and never keep any of their offspring, it probably wouldn't make any difference if subsequent generations are better than the first, but that rarely happens. From a purely economic perspective, it makes sense to get the best return on your investment, and there is very little — or nothing — to lose.

For a working example, we need only turn to the commercial dairy (cow) farmer. Almost invariably, these practical, tough-minded, cost-conscious farmers use the best purebred registered bulls or frozen semen they can find. They may not have the slightest intention of ever showing a cow or of raising registered cattle (although some of them are finding that registered cows are valuable for the same reasons registered bulls are practical). They use purebreds because it pays off in the milk pail. Milk production per cow has doubled since the last century. While some of this is due to feeding practices and other management details, a large share of the credit must go to genetics.

Similar progress is becoming the norm in commercial goat dairies, but there are still entirely too many half-pint milkers around that are being sold to unsuspecting novices who have heard that goats give a gallon of milk a day. The reason is simply poor culling practice — often starting with the selection of the buck. If you stay clear of a poor buck today, you save yourself the trouble of culling dozens or hundreds of poor does in the future.

Your chances of improving your herd are practically nil if you breed your does to a neighbor's nondescript pet buck simply because he happens to be cheap and available or if you buy a buck just because the price is right. On the other hand, if you choose a quality buck with a pedigree and records that speak to his genetic potential, you are making a positive impact on future generations in your herd. If you have paired him with a good doe and choose not to keep their female offspring, they can generate extra income as quality milkers for someone else. In short, you are becoming part of the solution instead of part of the problem. It is much better than foisting downgraded off-spring on a world that already has too low an opinion of these valuable animals.

Choosing a Buck

Whether you are choosing a buck to buy, to lease, or just for a visit with your does, the goal will be the same. So how do you choose a buck that will produce superior offspring?

With bucks, you generally get what you pay for. Good bucks are expensive, and seri-ous breeders won't sell any other kind. They're worth it, of course, because the buck you use this year will affect your herd for years to come. Even if you are leasing a buck, be suspicious of someone who is willing to provide breed-ing services for free. A quality buck is usually negotiated on a per-doe basis for small numbers and a flat fee for larger numbers. The price will depend on whether you are requesting service memos so you can register the offspring later.

You start by examining his pedigree, the record of his ancestors. If it's milk you want, make sure there's milk in that pedigree. While there admittedly are lovely grades that make milk by the ton, there is no way of knowing who their ancestors were or how good they were. A pedigree and milk-production records of several generations of forebears might not be insurance, but they're valuable management tools and much better than flying blind.

Next, consider conformation. If you have a doe with poor udder attachment or weak pasterns or any other fault that might affect her productivity and usefulness, you certainly won't want to breed her to a buck whose dam and granddam had the same faults. On the contrary, you want a buck that is particularly strong in those areas, so his daughters will be better than their mothers. Obviously, you can't see an udder on a buck, but you can ask to look at his mother, sisters, or daughters.

Not all goat owners will want to keep a buck, but his services will be needed periodically so your does will continue to produce milk.

Buck goats are much more masculine in appearance than does, as this Oberhasli demonstrates. Despite their powerful builds and disgusting (to humans) habits, bucks that have been well treated and trained can be quite docile.

Now suppose you select a fine buck of impeccable breeding, excellent health, and ideal conformation. Are your problems over? Not quite.

If you have four does — a fair average for a homestead herd — you can expect four doe kids the first year. Chances are you'll want to keep one or more of them. After all, didn't you buy the high-powered buck to improve your herd?

Now you have the problem of how to mate the daughters. While goats don't understand the term "incest" and will be just as happy to service close relatives as outsiders, it takes a lot of years of experience to really understand the fine points of genetic manipulation. In the hands of an expert, inbreeding may be the surest and fastest way to breed improvement. But it doesn't hurt to say this one more time: inbreeding emphasizes faults as well as good points. It's nothing to be dealt with haphazardly. (Actually, there is some evidence to suggest that inbreeding affects goats less than some other animals. But are your original goats good enough to be perpetuated — or should they be upgraded by outcrossing?)

Consequently, when your herd sire's first daughters come into heat, you'll want to find another buck.

Minimizing Faults

This is a good place to note that no animal is "perfect." All have faults of one kind or another, to a greater or lesser degree. It's the job of the breeder to eliminate those faults as much as possible in future generations, while at the same time preventing new ones from showing up.

An illustration of this would be a doe with very good milk production but a pendulous udder. That udder fault is going to shorten her productive life; it will make her more liable to encounter udder injury, mastitis, and other problems. So you'll want to breed such a doe to a buck that tends to throw daughters with extra-nice udders, in hopes that the offspring will have both good production and acceptable udders. Since both the dam and sire contribute to the offspring's genes, the second generation of udders probably won't be extra-nice, but they'll be an improvement over the dam.

The problem here is that, with four different goats in a small backyard dairy, there are likely to be at least four different faults! It's unlikely that even a good buck will be strong enough in four different areas to compensate for all of them. From the standpoint of breed improvement then, each doe in your barn is likely to be best matched by a different buck.

Living with a Buck

As already mentioned, bucks do smell, especially during breeding season. Girl goats (and some goat people) are inclined to like the musky aroma, but it will travel far beyond your barnyard; it will pervade your clothes, and even your living-room furniture will get to smelling like ripe billy goat, which for most people is less than desirable. Your neighbors might also have an opinion on the matter.

Also of interest to people who are new to goats is what they often call the buck's "objectionable, disgusting habits." Most city people are shocked when they find out that the cute and playful buck kid grows — astoundingly rapidly — into a male beast who not only tongues urine streams from females (and makes funny faces afterward) but who also sprays his own beard and forelegs with his own urine. This is natural goat behavior, but be that as it may, even many broad-minded people find it difficult to accept gracefully. Needless to say, the lovable buck kid loses a few friends when he reaches this stage.

Bucks are powerful animals — I've seen them snap 2 × 6s just for kicks — and one that has not been raised properly or finds himself in an untenable position can be a dangerous animal. I have never owned a buck that was any more hostile or aggressive toward humans than a doe, but they haven't been effeminate either, which would be a fault in a buck. Still, enough other people speak of "mean" or "dangerous" bucks, so it seems likely that they exist, and you should be forewarned.

Some women goat owners have said their breeding bucks tend to be inappropriately amorous toward them, perhaps because of some confusion of pheromones in the heat of the rut. Pound for pound, the buck probably has the advantage, so women should be aware that "bearding" the buck, or firmly grasping a handful of his beard, is a good emergency control method. It's probably not a bad thing for men to know about, too, but don't overuse it. Any animal will fight back if it is receiving hardhanded treatment.

Because they are powerful, and because of their natural sexual instincts, a buck requires much more elaborate and expensive housing than the does, especially during the breeding season. They must be housed separately, if only to avoid off-flavored milk, and an inadequately penned buck will soon be found with the girls.

While there are many advantages to buying a proven sire — a buck that is not sterile and who throws daughters with the traits you want in your herd — such bucks are either very expensive or old and otherwise worn out. Most bucks are sold as kids, fresh off their dam. Reputable breeders will take orders for their bucks in advance and either castrate or dispose of buck kids that are not spoken for. Half of all kids born are bucks, and only a small fraction of those are needed or good enough for breeding prospects.

The Buck Stops Here

Increasingly, many families aim to be self-sufficient in dairy products, and how self-sufficient can you be without a herd sire? If you start with fairly decent or above-average does and get a buck that's as good as or better than the does, it certainly isn't a catastrophe. But you should at least be aware of the information we've discussed here. If you're going to be a goat breeder, be the best you can, within the parameters you have set for yourself.

Artificial Insemination

There is another possibility that interests many people: artificial insemination, or AI. Along with embryo transfers and other genetic

---

DEALING WITH BUCK PERFUME

There is no getting around the fact that bucks in rut smell pretty bad. Not only that, but their odor also permeates skin, hair, clothing, and exposed fresh milk. Keep the buck away from the milking area and milking does except when breeding. A pair of slip-on coveralls and dedicated gloves work well when handling the boys. For some reason, goat-milk soap or shaving cream with aloe vera does a good job of getting the smell off skin.

technologies, AI is very popular with certain segments of the goat world. It's still not as common as in cows, but it has great potential for many reasons.

From the standpoint of breed improvement, there is nothing better. Anyone, anywhere, can use the finest bucks available and at low cost. In many cases one straw of semen (a straw is the thin plastic tube the semen is stored in) costs only a tenth of the same buck's standing stud fee, and you can use bucks that are thousands of miles away. Does can even be bred to bucks that are long dead; semen can be stored for years. Best of all, you can use proven sires — bucks that already have daughters, or even granddaughters, that are already milking and have production records. And the inbreeding problems previously mentioned in this chapter are easily eliminated with AI. Even goat keepers who aren't overly concerned about inbreeding can readily see the advantages of not having to keep a buck or having to traipse all over the countryside with lovesick does in the car.

At the same time, AI isn't the final answer to every goat owner's breeding concerns. In many cases it will be necessary for you to do the inseminating yourself. And although it's a relatively simple procedure, you won't learn how to do it by reading a book. Beginners typically have a 50 percent success rate. The best way to learn is by watching an experienced inseminator and asking plenty of questions. And you'll have to buy and maintain a liquid nitrogen semen-storage tank, so your breeding costs for just a few does will go up appreciably. Of course, you can always share those costs with a like-minded neighbor with goats or even a cow dairy owner nearby who will save a little space in a tank.

All of this is far beyond the basics of raising goats and the scope of this book. Look to the Internet or goat periodicals for ads for artificial insemination companies and contact them. Of course, you can also get the information you need to get started from goat raisers in your area who are using AI.

Caring for the Buck

Buck kids are raised very much like doe kids. They grow a little faster but take longer to fully mature. Yet even though a buck may not stop growing until he's 3 years old, he is capable of breeding a doe by the time he's 3 or 4 months old. Don't let his size or his kidlike demeanor fool you! Separate penning is necessary almost from the beginning and at least by 2 months of age.

Housing

A buck's housing must be especially sturdy. It's difficult to "overbuild" a buck pen! The barn or shed can be simple, but it should have walls of 2 × 6s or 2 × 8s or cement blocks. Walls, pens, and fences must be strong enough to withstand the assault of a 200-pound (90 kg) battering ram with a head as hard as a sledgehammer. A shed of 6 feet by 8 feet (1.75 by 2.5 m) will provide sufficient shelter, with an exercise yard at least 6 feet by 30 feet (1.75 by 9 m). Posts — extremely sturdy and well set in the ground — should be no more than 8 feet apart. While typical fencing material is 4 or 5 feet (1.25 or 1.5 m) high, it's not likely

to slow down an amorous buck. Some bucks have been known to scale a 6-foot fence to get to the object of their affection, but they have to work pretty hard to do it. Don't make it easier by locating buildings, climbing toys, or even tree limbs close to the fence.

Feeding

Feeding a buck is less complicated than feeding a doe, since the male needs less protein and minerals. They can get the same feed, with two exceptions: don't feed bucks pure alfalfa or large amounts of grain. Both can cause imbalance in the phosphorus-to-calcium ratio that is critical to urinary tract health. A diet that is too heavy in phosphorus causes tiny stones of calcium to form in the urine. Anyone who has dealt with kidney stones knows how painful something like that can be. For a buck with a very narrow urethra, an obstruction can cause urine to back up, and the buck can die a slow and very painful death. Feed bucks a good-quality grass hay.

Training

Early training pays off. Teach a buck kid to lead with a halter or strong collar while he's still small enough for you to handle. Some people teach bucks to lead by the ear or with finger pressure under the jaw. Use patience and persistence, and he'll remember, which can be very helpful when he weighs over 200 pounds (90 kg) and could easily overpower you.

Don't let a kid do what you wouldn't want the adult to do. If you think it's cute to have his little feet on your lap as a kid, imagine those feet with 200 pounds of body behind them. Training a young goat not to jump up on you is similar to training a dog. Firmly lift your knee to give your goat a sharp jab in the ribs while

BUCK BARN AND YARD

The buck barn and yard should be sturdily constructed. This one is attractive and serviceable, but the herd sire would enjoy it more if it were less square. The ideal yard can be as narrow as 6 feet (1.75 m) wide, but it should be at least 30 feet (9 m) long.

Another improvement would be a means of feeding and watering the buck without entering the pen, and a much higher fence that will hold him during the breeding season. During the rut, a buck will scale, hurdle, punch a hole through, or otherwise thwart any fence that isn't seriously constructed.

saying "no" in an authoritative voice. He'll learn amazingly fast. The object is not to hurt the animal but to be firm enough that it is an unpleasant experience.

Again, the size and strength of an adult buck can be dangerous when does are in heat. Any male of this size is to be handled with respect and caution.

When and How Often to Breed

A buck can be used for limited service even before he's a year old. Most authorities say he should be limited to 10 to 12 does his first year. A mature buck can service more than 100 does a year, and of course, with artificial insemination, the number can be considerably larger than that.

10

BREEDING

A goat obviously must be bred in order to produce kids — and milk — and this chapter will give you a good feel for what you need to know to help both you and the doe get through the event with the least amount of trauma. A doe is typically bred every year, as she would normally be in nature, but it doesn't necessarily have to be that way. Before jumping into the birds and bees (and bucks), let's look at an alternative.

Milking Through

A doe is ordinarily bred once a year to produce milk to feed her kids. In theory, she needs to give birth on a regular basis in order to produce a reasonable quantity of milk. This can be seen in the lactation curve: the amount of milk produced daily increases rapidly after parturition, and then slowly drops off, reflecting the natural milk requirements of the young.

A 305-day lactation is considered standard but not because it's the norm. Milking records are based on a 305-day lactation, assuming 365 days between birthings and a 60-day dry period for the doe to rest and replenish her body reserves. Many goats don't even have the genetic capacity to milk a whole 305 days and are dry much earlier.

On the other end of the scale are the goats that are "milking through." This refers to does that continue to produce respectable quantities of milk for 2 or 3 years or even more, without being rebred. For many home dairies, this is an attractive option.

The main problem is finding a doe that is capable of such production. If we assume that most beginners will start with an average animal, the chance that your first goat will be capable of milking through is pretty slim.

How will you know if you have a goat with this potential? You will have a pretty good indication if she continues to produce well even when she would normally be bred. How much milk is "enough" is up to you, but most home dairies would be satisfied with 4 to 5 pounds (1.75 to 2.25 kg) a day. While a commercial dairy needs a certain amount of milk per doe per day to be profitable, a home dairy might be satisfied with just a few glasses a day.

The Advantages

A family goat that milks through offers many advantages.

One is that you'll have milk year-round. The "normal" system of drying off after 305 days leaves you with the unpleasant situation

---

This crossbred Saanen doe is late in her pregnancy, when older females carrying several kids can look as wide as they are tall. Because of her heavy winter hair coat, a modest dairy clip would be in order before she kids.

of buying feed for the goat and milk for your family for 2 months of the year! The option is having several goats freshening at staggered intervals, which, among other things, means having a flood of milk at certain times. It isn't always convenient and often involves winter lighting schemes or progesterone hormone implants to stimulate ovulation out of season.

Another big advantage for many people with just a few milkers is that they don't need a buck or don't have to transport the does for breeding. This reduces costs, time, and labor.

A third advantage won't apply to as many goat keepers as the first two and might not be as readily apparent, but it can be significant. That is unwanted kids. For many people, kids are the most delightful part of raising goats.

Why would anybody not want them? And on a homestead, if they're not needed as milkers, the meat is an important by-product of the dairy enterprise.

But raising kids requires a great deal of time and effort. There is also the expense of milk lost to the household or dollars paid out for milk replacer. And let's face it, many goat raisers can't even think of butchering and eating kids. But if your little herd doubles or triples in size every year, something has to give. (Do the math: If a doe has doe twins once a year, and her kids and their kids have doe twins, you'll go from 1 to 3 to 9 to 27 to 81 goats in just 5 years! Not all kids are females, naturally, but you will get "compound interest" from your goats.) You can't keep them all. And if disposing of them is a problem, milking through is an elegant solution.

Other reasons for milking through are more subtle. One is that kidding complications, such as ketosis, are a leading cause of death for high-producing does. When kidding is reduced to every 2 or 3 years, the potential danger is eliminated.

Remember, not all goats are persistent milkers. Saanens are most likely to milk through, but individuals of other breeds will, too, and of course not all Saanens will. Like many other management practices, milking through depends on your specific animals and your own needs and preferences.

There is one caution with deciding to milk through. If you make the decision to milk through, and then your doe slows down

A few goats can be milked by hand, but a milking machine takes a lot of work out of the twice-daily operation with a larger herd. Silicone teat cups aren't as heavy as stainless steel used for cows.

to a dribble in midsummer, she is not likely to come into heat again until fall. You will not only be without milk for the remainder of the summer, but you will also need to wait without milk an additional 5 months after you get her bred.

Preparing for Breeding

So you've decided it's time to play cupid. It all starts with the doe's estrous cycle or heat periods. Goats and most other animals "cycle"; that is, they are fertile only for relatively short periods at more or less regular intervals. Unlike cows, or hogs, or rabbits, which come into heat year-round, goats generally come into heat only in the fall and early winter. A doe usually will accept service from a buck only when she is in standing heat. If the time isn't right, she will ignore or avoid his attentions. At that point, copulation won't result in pregnancy anyway, because the sperm and the ova aren't in the right place at the right time.

Seasonal breeding has decided advantages for animals such as deer and goats in the wild. Their young are born when it isn't too cold; there is plenty of lush, milk-producing feed for the mother and tender grasses and leaves for the young to be weaned on; and the offspring are fairly strong and independent by the time the weather turns harsh again. Desirable as such an arrangement may be for wild animals, it puts the dairy goat farmer in a bind.

Lactation and Seasonal Curves

In chapter 2 we examined the lactation curve. If you plot such curves for several does, all of which have been bred at more or less the same time, it's apparent that the goat farmer will be swimming in milk during part of the year and dry as a bone in another part. The normal lactation curve is reinforced by seasonal curves that are equally normal in both cows and goats, due to feeding conditions, weather, extended hours of sunlight, and other factors.

Animals simply produce more milk in summer than in winter. This is perhaps the single most serious drawback for commercial goat-milk producers. Cheese and fluid-milk processing plants need to have raw material coming in the door year-round. In many cases, they actually put maximum production ceilings on summer milk and pay a premium for winter milk because it is so much harder to come by. It's easy to see that the poor goat farmer has a problem.

Backyard dairy operators and homesteaders share the same dilemma but to a smaller degree. If you have just one goat, even if she has a lactation of 10 months, you'll have plenty part of the year and be without any milk at all for 2 months of the year. With two goats you can attempt to breed one in September and one in December. Then, theoretically, you will never be without milk, but a look at the lactation curves plotted together will show that your milk supply will be far from steady. You'll have too little or too much far more often than you'll have just enough.

The point here is that you'll want to have your does bred as far apart as possible, while avoiding the risk of having a doe miss being bred at all. With some does, in some years, even December may be pushing it; they simply won't come in heat again until September. There are kids born in every month of the year, but as a practical matter for the small-scale raiser, you can't count on out-of-season breedings.

Detecting Heat

For many beginners, and especially those with only one or two goats, it's very difficult to tell when a goat is in heat. The usual signs are increased tail wagging; nervous bleating; a slightly swollen vulva, sometimes accompanied by a discharge; riding other goats or being ridden by them; and, sometimes, by lack of appetite and a drop in milk production.

If a buck is nearby, there will be no doubt: she'll moon around the side of her yard near the buck pen like a lovesick teenager, and the buck will probably respond by trying every trick in the book to get to her side of the fence.

If you lack a buck and have trouble detecting heat periods, or just want to make very sure she's in heat before you make a several-hour trip to a buck, you might use this trick. Rub down an aromatic buck with a cloth or tie one around his neck for a couple of hours. Seal it

tightly in a clean cottage cheese or other disposable plastic container. When you suspect your doe is eager for male companionship, give her a whiff of that cloth, and your suspicions will quickly be confirmed or denied. If she shows little more than mild interest, she's probably not in heat. If she wags her tail furiously and tries to climb in the container with the rag, it's time for action.

But you can't breed a doe with canned buck aroma. If you don't have a buck, you'll have to take her to one.

Transporting the Doe

When we had neither a buck nor a pickup truck, we used to transport does in the closed trunk of the car. Of course, 1965 Pontiac

---

LEASING A BUCK

Some owners will lease their buck for breeding when he is finished with work at home. For your own financial safety — since some bucks can be worth several hundreds or thousands of dollars — have a lease agreement that stipulates who is responsible for the welfare of the buck and what happens if he is injured or dies. Most reputable breeders understand that accidents happen, but there are a few of those breeders that give the industry a bad name by passing along unhealthy breeding stock and then taking it out on an unsuspecting goat owner if the goat finally gives up the ghost on your doorstep. When leasing a buck, you will want to keep him through two heat cycles or 42 days to be sure your doe does not come back into heat.

HOW TO PLAN AHEAD

Keep a calendar record of when a doe is in heat. By noting the number of days between each heat cycle, you can calculate pretty accurately when she will next cycle and be ready for the buck. See Gestation Chart, page 268, for more information.

trunks were more spacious than some small pickup trucks today, and this is no longer an option. There's no way I can afford a livestock trailer, and a pickup truck, even with side racks, is far more hazardous for the goat.

If your car isn't too fancy or if you really love goats, she can ride in the back seat. If she's lying down, she will not "disgrace herself," as one puritanical old goat book put it. Some goats tend to get carsick standing up and will be too woozy to be interested in breeding after the trip. On the other hand, some enjoy riding in a car as much as dogs do, even to the point of sticking their heads out the window. It's very entertaining to watch the reaction of other drivers who have never seen a goat riding in a car.

Successful Breeding

A doe will be in standing heat for 24 hours, although this varies widely. If she is not bred, she will come into heat again in 21 days; this is also an average that varies considerably.

If a doe is serviced and still comes back in heat, there could be several causes. She might not have been bred at the most opportune time. Maybe one more try will do the trick. (It isn't necessary to leave the buck and doe together for long periods. If the doe is really in standing heat, one service is sufficient, and that won't take more than a minute, which sometimes seems silly after you've spent

an hour on the road and still have to drive home again!)

Breeding Problems

Sterile bucks are rare, and if a buck is sterile, obviously, none of the does he serves will conceive. However, sperm can have reduced viability at certain times due to overuse of the buck, recent deworming with an injectable dewormer, and other factors.

If a doe simply will not get bred, the most common cause is cystic ovaries. Cystic ovaries are usually obvious, because the doe repeatedly comes into heat every 10 or fewer days and won't stand for breeding. The problem can usually be treated by a veterinarian with a couple of hormone shots, but it will mess up a schedule if you wanted her bred for a specific kidding date.

Overly fat does are often difficult breeders because of a buildup of fat around the ovaries.

Another serious condition, although it's not as common as we once thought it was, is hermaphroditism, or bisexualism. For somewhat technical reasons the term "intersex" is now preferred in some circles as being more accurate, but "hermaphrodite" is easier to use and more colorful.

The hermaphrodite looks like a doe externally, but it actually has male organs internally. Not all have obvious external abnormalities. Carefully examine the vulvas of newborn kids.

THE FIRST HERMAPHRODITE

The term "hermaphrodite" comes from Hermaphroditus, the son of the Greek gods Hermes and Aphrodite; his name is a combination of theirs. According to myth, the nymph Salmacis fell in love with Hermaphroditus, who was bathing in her fountain. She wanted the two of them to achieve complete union, and her wish was literally granted.

Press gently on each side of the vulva and watch for a protrusion about the size of a pea at the bottom of the vaginal slit. "Bucky" behavior in a normal-appearing doe kid is cause for suspicion, as is a bulky, male-looking head and shoulders. Intersex goats are often either overly aggressive or unusually withdrawn.

In goats the condition is often related to the mating of two naturally hornless animals (obviously not referring to animals that were disbudded or had their horns removed surgically). The genetics get a little complicated, but basically you must determine whether a naturally hornless buck is homozygous or heterozygous; that is, whether he inherited a gene for horns from either of his parents or did not. Theoretically, there can be no homozygous does because they'd be hermaphrodites and couldn't have offspring. Both types of bucks will produce some hermaphrodites when bred to hornless does, according to theory, but the homozygous hornless bucks will produce more.

If either the buck's sire or his dam were horned, he's heterozygous. If neither parent was horned, you can't be sure what he

is without seeing a number of his kids. If any of the kids have horns, the buck is heterozygous. If all the kids are hornless, even out of horned does, chances are the buck is homozygous.

All of this is of great interest to geneticists, large goat breeders, and people who take a keen interest in breeding, but the average backyarder or homesteader is better off to follow the lead of the major commercial goat farms and just avoid hornless-to-hornless matings.

When to Breed

Doelings are sexually mature as early as 3 or 4 months of age. They should not be bred at that stage of development. In most cases spring kids that are well developed and healthy should be bred when they weigh 70 to 80 pounds (30 to 35 kg), which comes at 7 or 8 months old. That means they'll kid at 1 year of age.

Breed by weight, not by age. Being bred too early will adversely affect a doeling's growth and milk production. But being bred later doesn't contribute to their health or welfare or your finances. It's expensive to keep dry yearlings, and records show that does that kid at 1 year of age produce more milk in a lifetime than those that are held over. Kids that are held over before breeding also tend to get fat, which causes breeding problems. Many people mistakenly hold back young does because "they look so small" or because 7 months seems so young. With proper nutrition, they'll produce healthy kids and will keep on growing themselves.

Drying Off

The older doe will be, or should be, still milking when she's bred, but advancing pregnancy will cause most does to dry off. Some people who really want milk will continue milking a naturally drying-off doe as long as she's giving a few squirts; others figure even a pound isn't worth their trouble, so they stop milking, and the doe dries up.

In any event it's a good practice to dry off a doe 2 months before her kids are due. In fact, studies support the need for at least 60 days for the mammary to shut down and then rebuild itself for another round of milk production. Growing unborn kids puts tremendous demand on the doe's body in the last few months of gestation. If she is also producing milk, it could compromise her health and that of the kids.

In most cases simply quitting milking and cutting off the grain ration for a period will cause the animal to dry off naturally. In cases of extremely heavy or persistent milkers, it may be necessary to milk her out at increasingly longer intervals (by the way, this is the kind of goat that might make a good candidate for milking through). Some people milk once a day for a few days, and then every other day, and then stop. This sends mixed signals to the mammary. It's better to cut off the grain for a few days and then just stop milking, period. The doe may be uncomfortable for a few days, but the pressure on an engorged udder pushes a physiological "off button" that causes the udder to reabsorb milk and stop producing.

Feeding the Pregnant Doe

No good dairy animal, cow or goat, can eat enough during lactation to support herself and her production. That's why she requires a rest to build up her body before peeling off her own reserves to fill your milk pail. It's been said that for each pound (0.5 kg) of increase in body condition during the dry period, a Holstein cow will produce an extra 25 pounds (11.5 kg) of milk, a Guernsey an extra 20 pounds (9 kg), and a Jersey an extra 15 (6.75 kg). We could anticipate proportional results with goats.

This is not to suggest that a pregnant animal should be overconditioned or fat. A dairy animal should never be fat, just in good condition. In fact, fat causes problems in

HOW TO DRY OFF A DOE

The best way to dry off a doe is to stop milking her.

1. At the last milking, use a dry-cow antibiotic infusion to reduce the possibility that mastitis will develop during the dry period (a teat is infused by injecting an antibiotic into the teat canal). This is an extra-label use of antibiotics, and withdrawal times have not been calculated for goats, so consult with your veterinarian first. Also use a teat dip (see chapter 8).

2. Stop feeding grain. Once she is dry, you can switch to a dry doe ration, which has reduced concentrates and lower protein content.

3. Dip the teats twice a day for the first 4 to 5 days. Pressure builds in the udder and signals the body to stop producing milk.

4. If there is still substantial pressure in the udder after 4 days, milk out the doe, reinfuse the udder, and start the procedure over again. Dispose of the antibiotic milk.

breeding and pregnancy. But the goat needs a well-balanced diet. Too much feed produces kids that are too large to be easily delivered. Excess minerals in the doe's diet produce kids with too-solid bones, which also causes difficulty.

For the first 3 months of pregnancy, an older doe is probably still producing milk for your table. Continue to feed her on a milking diet (1 pound [0.45 kg] of grain per 2 pounds [1 kg] of milk produced), but be sure she gets plenty of good-quality hay, because her body is working doubly hard. You can drop the grain for a few weeks while she is being dried off, but then start her on a dry-goat ration of about 12 percent protein. The first 3 months for a doeling also requires a fibrous diet and rather low protein.

Most of the kids' growth comes in the last 8 weeks of pregnancy. During this period the ration should be changed gradually, not only because almost three-quarters of the kids' growth is taking place, but because this is when the doe needs to build up her own reserves for her next lactation. The next few months are kind of a balancing act. If the doe is underweight, she should be fed ½ to 1½ pounds (0.25 to 0.75 kg) of 12 to 16 percent protein grain a day. The protein will depend on the quality of hay she is getting. If it's alfalfa, use the lower number. If it's grassy, use the higher protein. At this stage free-choice minerals are especially important, especially vitamins A and D, iodine, and calcium. Beet pulp or bran will provide required bulk, and molasses will supply some iron as well as the sugar that helps prevent ketosis, while having a desirable laxative effect.

If the doe is overweight, it is important to get the vitamins and minerals in her, but a good mixed hay and minimum quantities of grain at a 12 percent protein level are fine.

Finally, everything is ready. The goat stork cometh.

Goats are herd animals and will be happier if left in the company of other goats during their pregnancy. If you don't have a separate pen available for kidding, she will find a clean, quiet corner where she can go about the business of motherhood.

11

KIDDING

The "miracle of birth" is aptly named. Like all miracles, it's invested with wonder, awe, excitement, and joy. There have been cases of people who would have nothing to do with goats — until they saw newborn kids frolicking in fresh clean straw and fell in love. (I'm married to one of these people.)

There is little doubt that the first kidding season brings the new-kid excitement that is hard to duplicate in today's electronic and plastic world. Most new goat keepers, judging from the mail I get and my own first experiences, are scared silly as parturition approaches. A large part of this fear comes, I believe, from reading books and articles describing all the things that can go wrong. You expect the worst. But goats have been having kids all by themselves for thousands of years. While problems are certainly possible, 95 percent of all goat births are completely normal and won't even require your assistance. The chances for a normal birth are enhanced by proper feeding and management, especially during the latter stages of pregnancy.

Anticipating the Delivery

The average gestation period for goats is 145 to 155 days. There is a tendency for does with triplets to kid slightly earlier than does with single kids, but both are usually within this time frame. Some experts say there is evidence that goats and sheep can control the time of birth to coincide with copacetic weather conditions. Other people say they control it, all right, but usually to have the kids and lambs arrive on the coldest, most miserable night of the year or else while you are away. Either way, many goats seem to kid at about the same time with every freshening. Record that time for each doe, and next year you might be forewarned.

KIDDING SUPPLIES TO HAVE ON HAND

- 7 percent iodine and dipper (note that this is not the same as iodine-based teat dip, but the dispenser used for teat dipping is handy for this job too)

- Clean cloth towels

- Small, clean water pail

- Feeding tube (lamb size) and syringe

- Alfalfa hay

- Nipples and bottles

- Colostrom (frozen)

Within minutes of entering the world, the newborn kid will struggle to its feet and search for its first meal.

Checking for the Signs

Start checking your animals frequently and carefully 140 days after breeding. When a doe is getting ready to kid, she will become nervous and will appear hollow in the flank and on either side of the tail. There may be a discharge of mucus, but this can appear several days before kidding. When a more opaque, yellow, gelatinous discharge begins, it's for real.

Kids can be felt on the right side of the doe. It's good practice to feel for them at least twice a day. As long as you can feel them, they won't be born for at least 12 hours. Also, if you feel the doe regularly you'll be able to notice the tensing of the womb. After this, one of the kids is forced up into the neck of the womb, causing the bulge on the right side to subside somewhat. This will be noticeable only if you have paid close attention to the doe in the days and weeks before. The movement of the first kid also causes the slope of the rump to move into a more horizontal position. At this point you can expect the first kid within a couple of hours.

Many people look to "bagging up" or enlarging of the udder as a sign of approaching parturition. This is unreliable. Some goats don't bag up until after kidding, and others will have a heavy milk flow far in advance. In some cases, if the udder becomes hard and tight, it might be necessary to milk out the animal even before kidding.

A better indication is a softening of the ligaments from the tailhead to the pinbone. Check these earlier so you know how they normally feel, which is like a pair of pencils. As the doe gets within 48 hours of kidding, the two ligaments will go from pencils to thick rubber bands and then to bubble gum. When they seem to have disappeared, the doe should kid within 12 hours.

Preparing the Facilities

Although we humans try to take good care of our animals, we often complicate things. We've mentioned feeding of the pregnant goat, which can affect the ease with which she delivers. A free-ranging, experienced goat knows what to eat, but if we are the ones bringing her food, she must depend on our judgment. Likewise, the goat kidding outside on her native mountain range knows what to do when her time approaches. She is probably safer and in more hygienic surroundings on her mountain than she would be in your barn. It's just about impossible to duplicate natural conditions for domestic animals.

There are innumerable instances of goat owners going to the barn for morning chores and finding a couple of dried-off, vigorous, and playful kids in the pen with their mother. But it's definitely preferable to have some idea that the kids are on the way and to make certain preparations for them.

The doe should have an individual stall for kidding. It should be as antiseptic as possible and well bedded with fresh, clean litter. Something softer than long straw, such as chopped straw, is preferable if you have it.

Don't leave a water bucket in the kidding pen. If possible, leave it where she can reach it but not accidentally drop a baby in it. For the most part, she isn't going to be interested in drinking at this point anyway.

The Birth

One study of owners' records shows that 95 percent of goat births are uneventful. For normal births, your assistance will be unnecessary and perhaps even unwanted. During labor you'll have nothing to do: you don't even have to boil water, unless you want to make a cup of tea.

The Three Stages of Kidding

Kidding normally occurs in three stages. The first stage is when contractions of the uterus force the placenta, fetus, and fluids against the cervix, dilating it. This can last up to 12 hours in first fresheners, but older does are usually faster. Second-stage labor involves "straining," or contraction of abdominal muscles. This typically lasts about 2 hours or less and ends with the expulsion of the last kid. The third stage involves the expulsion of the placenta, or afterbirth. Four hours is considered the norm for this, although it can take several hours more; much longer than that and it's a retained placenta, which calls for the services of a veterinarian.

Assisting the Delivery

If the doe has been pushing for a while and seems concerned and nothing is happening, you'll have to help. Remove rings, watches, and long fingernails. Insert your disinfected, lubricated hand and arm into the birth canal to find out what's wrong. (If there is still a tight ring of tissue a few inches inside the vagina, the cervix is not dilated far enough, and you're jumping the gun, but the cervix is usually fully dilated before the pushing stage begins.) A germicidal soap will serve as disinfectant; mineral oil or K-Y Jelly is an acceptable lubricant.

If you've never seen a newborn kid, this is not only scary, it's difficult to imagine what you're feeling for. Try to sort out the heads and legs, and if necessary, rearrange them in the proper presentation position. In most cases, it will be a simple matter to "lead" the first one out the next time the doe strains. Pull, but very gently, working with the doe; otherwise, hemorrhage might result. Chances are the others will come by themselves soon after.

If you need to do manual exploration, an antibiotic bolus, available from your veterinarian, should be inserted into the vagina after the last kid and placenta are delivered.

Goats usually have two kids, but three, four, and even five aren't all that uncommon. If no more come within half an hour and the mother seems relaxed and comfortable, you can assume that's all there are.

If you happen to encounter a difficult birth, a doe struggling to expel a dead kid for instance, get a veterinarian or experienced neighbor to help. But again, such help is seldom needed. Be aware of what can go wrong, but please don't make yourself sick worrying about it beforehand. Being anxious about an imminent birth is normal, but remember that 95 out of 100 births will go smoothly without any assistance. Relish the experience!

A new kid gets a thorough tongue washing from its attentive mother, but sometimes she'll start at the wrong end. If you have attended the birth, be sure the kid's nose and mouth are clear of the placental sack.

THE KIDDING PROCESS

---

1. When a doe gets down to serious business and is straining to push her kid out, the first thing to appear is often a "bubble" of thin tissue filled with yellowish amniotic fluid. Sometimes feet and even a nose will appear in the bubble before it bursts naturally.

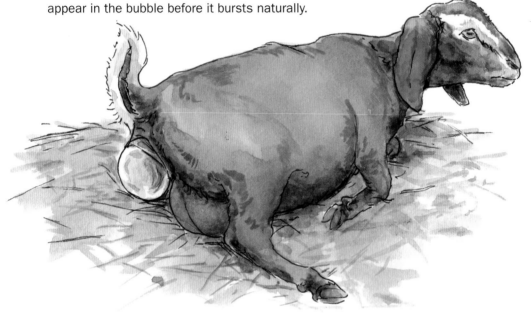

2. A normal presentation is two little feet pointed down and a nose between them. At this point, the softened tissue around the vagina stretches to allow the head through the opening. With a few good pushes, the narrower body will follow.

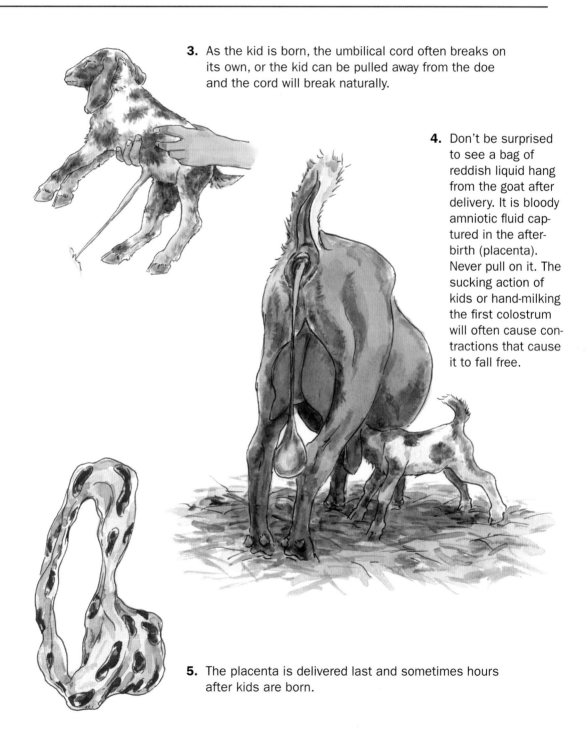

3. As the kid is born, the umbilical cord often breaks on its own, or the kid can be pulled away from the doe and the cord will break naturally.

4. Don't be surprised to see a bag of reddish liquid hang from the goat after delivery. It is bloody amniotic fluid captured in the after-birth (placenta). Never pull on it. The sucking action of kids or hand-milking the first colostrum will often cause contractions that cause it to fall free.

5. The placenta is delivered last and sometimes hours after kids are born.

KID PRESENTATIONS

Normal. The kid's nose is positioned between the front feet. Picture it in your mind's eye, and you'll see that this presents a more or less cone or wedge shape that gradually distends the vagina and makes the birth easy.

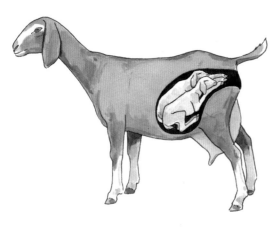

Multiple Births. Twins are more common than single births, and triplets, or even quadruplets, are not rare. In some multiple births, the kids and umbilical cords get all tangled up inside the womb. If the doe is pushing with no success, use your lubricated hand to follow the legs down to the shoulder and make sure both presenting legs and nose all belong to the same kid.

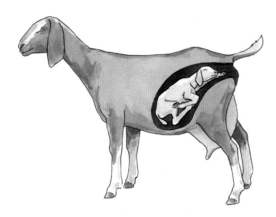

Foreleg or Both Legs Back. One leg or both are bent back, the wedge shape is missing, and the bent leg might complicate the birth. If the doe is having difficulty, it might be necessary for the caretaker to reach in and bring both legs forward.

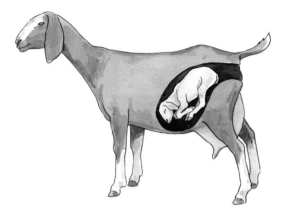

Breech. A breech delivery inverts the wedge shape and can cause a difficult birth. The biggest danger is that if the sac breaks before the kid is completely out, it could suffocate. Turning a breech kid all the way around is a major job and can cause some serious tangles. Sometimes it's best to let the kid lead with the two hind legs and then help deliver the head as fast as possible.

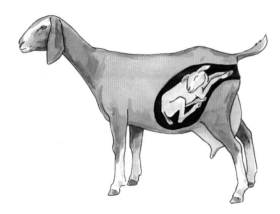

Head Turned Back. If the kid's head is turned back, it might have difficulty entering the birth canal. The solution is to reach in and move the head so the nose is between the two front feet.

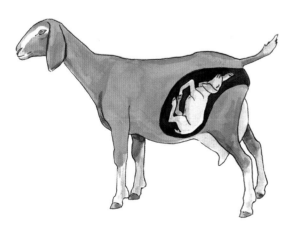

Upside Down with One Leg Back. In this unusual position the kid is upside down and has one front leg back. If the doe is having difficulty, the caretaker will have to go in and rearrange the kid for a normal presentation.

The Umbilical Cord and Afterbirth

In most cases the umbilical cord breaks by itself, or the doe severs it with a few bites. Occasionally, the remaining cord is very long and can get wrapped around legs and other hazards. The fancy way to handle it is to tie the cord shut with a piece of string about 3 inches from the kid's body and then cut the cord on the doe's side. Less fancy, but just as effective, is to pinch the umbilical cord near the kid's navel and cut the cord about 3 inches out. It takes about a minute of pressure to be sure the blood has stopped pumping through the cord, and then dip it in iodine.

New mothers don't always understand that the head is the end to start cleaning, and the kid might not be able to breathe through any amniotic fluid or parts of the fetal sack still clinging around its head and mouth. One of the most important accessories of the midwife is a good supply of clean cloths, and lots of them! Wipe the kid off, paying special attention to make sure the mouth and nostrils are clear of mucus. The doe will gladly help by licking her newborn. If you are on a caprine arthritis encephalitis (CAE) or CL prevention program, of course, it will be up to you to make sure that the doe does not lick the kid (see page 127). In cold climates, wet feet and ears freeze quickly. To prevent them from developing frostbite and eventually losing tissue, be sure the ears are dry and fluffy before leaving the kid on its own. (Unfortunately, some does will show their maternal instincts by continuing to clean the kid's ears until they are wet again, and they subsequently freeze. Keep watch.)

The umbilical cord must be disinfected to prevent bacteria from traveling up it into the kid's body. Iodine spray is convenient, but better protection can be had by pouring 7 percent iodine into a small container, pressing it up to the kid's navel, and then briefly tipping the kid over to ensure good coverage of the entire navel with the iodine. The umbilical cord will dry up within a day and fall off by the end of a week.

Watch for the afterbirth. Normally, the action of a kid suckling or the owner milking the doe will stimulate the uterus to contract and expel the placenta. If the doe doesn't expel it within several hours, call a veterinarian; a retained afterbirth is nothing to mess around with. If it's just hanging out of the doe, don't pull on it. You might cause hemorrhaging. It can continue to hang for as much as 24 hours, but that is rare.

When the afterbirth is expelled, dispose of it. Some does eat it, a natural instinct most wild animals have developed to protect their young from predators that might otherwise be attracted. Most goat keepers don't care for this little ritual, but it will neither help nor hurt the goat.

Attending to the Doe

Put the kid in a clean, dry, draft-free place (a large box works well), and turn your attention to the doe. She has lost a tremendous amount of heat, even in warm weather. Offer her a bucket of warm water. Some does seem to appreciate the water even more if a cup of cider vinegar or molasses is added.

Most pet-type goats also get a special treat at this point, perhaps a small portion of warm bran mash or oatmeal or a handful of raisins. Provide some of the best hay you have — she's earned it.

The doe will continue to drip dark brown blood and vaginal fluid for a few days and will then seem to clean up. If the discharge takes on a foul smell, there may be a uterine infection present, and the help of a veterinarian is called for. Don't be upset a week or so after kidding if you see a sudden discharge of brown glop for a few days. It is part of the cleaning process as the uterus slowly returns to normal size. Again, if it doesn't smell foul, don't worry about it.

Caring for the Newborns

Now it's time to attend to the babies. Following are your responsibilities.

Kid Inspection

Examine the kids. Many people who don't want to bother raising buck kids for meat euthanize them at birth. (The easiest way — if there is anything easy about euthanizing — is to immediately lower them into a bucket of warm water, where they will drown. They have just come from a warm, wet environment and don't seem frightened by a return to the same place.) Others who can't bear to do this try to give them away to people interested in keeping a buck as a pet, a companion animal for a horse, or as a harness or pack goat. And remember, only the very best bucks from outstanding dams and sires should be kept for breeding. One mature buck can breed a hundred does a year, and many bucks are kept for 5 years or more. Mathematically, this means that less than one in five hundred is needed as a sire. It's highly unlikely to be the one just born in your barn.

On doe kids, check for supernumerary (extra) teats. There are several variations of this condition, some of which make the animal worthless as a milker. If the extra teat is sufficiently separated from the main two, it might not interfere with milking and can even be removed at birth with surgical scissors or by tying a fine thread very tightly around the base and letting it atrophy. Actual double teats make the animal worthless. Bucks can also have double teats, and such animals should not be used for breeding.

If the tip of the vulva on doe kids has an obvious pealike growth, the animal is a hermaphrodite (see page 157) and will not breed. It should be destroyed, and the mating that produced it should not be repeated. Not all hermaphrodites, however, will display this growth, and very, very few goat owners will ever see a hermaphrodite in their herd unless they do a lot of hornless-to-hornless breedings.

Unwanted or cull kids, both bucks and does, can be raised for meat for your own table or for others'. If you have a better use for the milk and don't want to bother with milk replacer and the work of hand-raising kids, they can be butchered at birth and dressed like rabbits. Wait until they dry off; then they're easier to handle (see chapter 15 for more on butchering and meat). If you are going to raise meat kids, there's no reason you can't leave the kids on their dam. Dan Considine, who has been raising goats for more than 45 years, calculated the amount of milk that the meat kids consumed and the amount that went into his commercial bulk tank and concluded that pound for pound (and dollar for dollar) he was ahead financially by leaving the kids on their dams.

Keeping the Kids Warm

Perhaps one of the most common kidding problems people encounter is entering the barn on a blustery morning in late winter or early spring to find a newborn kid cold and shivering. If it seems to be doing all right, don't feel sorry for it and bring it in out of the cold. If the mother is going to be nursing it, make sure the kid is dry and set it in a draft-free corner where Mom can go on with her part of the job. If you are going to be raising the kid by hand, place it in an enclosed box or pen padded with an old blanket or feed sacks, away from any hint of a draft, and with a heat lamp if the weather is really nasty. But don't let it get hot. A switch back to normal temperatures will be as dangerous as the cold that brought on the problem in the first place.

In the case of a severely chilled kid on the brink of death, more drastic action will be required. If you find one still wet and thoroughly chilled and nearly lifeless, one way to save it is to submerge it up to the nose in a bucket of water at around 105°F (40°C), which is about the temperature of the environment it just came from. When it has revived, dry it thoroughly, wrap it in a feed sack or blanket, put it in a box in a protected place, and watch it carefully.

However, such a kid might also be suffering from hypoglycemia, or low blood sugar.

Fresh straw, a solid wall, and heat lamps all provide the newborns with a good start to life. Note that each kid is wearing a disposable ID collar printed with birth information. After tattooing or tagging, the collar can be removed.

This is a cutaway view of a warming box. Note the hinges at the top, on the wall. Such a box can keep kids more comfortable in extremely cold or drafty weather, but newborn goats can stand surprisingly low temperatures if they're in a draft-free place. Try not to have more than four or five kids, or kids of different ages, in a box. In their attempt to stay cozy, it is not unusual for one — usually the smallest — to be smothered in a kid pile.

As the glucose level falls, the kid shivers and arches its back, its hair stands on end, and it moves stiffly. Eventually, it lies down, curls up, becomes comatose, and dies.

The remedy is to warm the kid thoroughly and administer at least 25 milligrams of 5 percent glucose solution with a small rubber stomach tube. When the kid is showing signs of reviving, get 2 ounces (60 cc) of colostrum into it, with the stomach tube if necessary. Never put anything into the stomach of a goat with a core temperature below normal. Return the kid to the barn as soon as it's active.

If you do end up taking a kid into the house for any reason during cold weather (and almost everyone does), you're stuck with a goat in your

HOW TO USE A STOMACH TUBE

A kid that is too weak or comatose to suck must be fed with a stomach tube. (Stomach tubes are available from sheep- and lamb-supply houses, though a sterile catheter from a hospital or nursing home will work.) The stomach tube is a small, flexible plastic tube to which you attach a 60-mL syringe. The kid must not be fed if its gut is still cold.

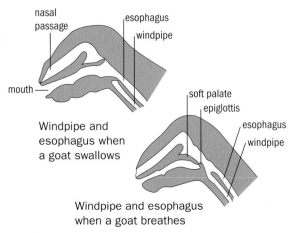

Windpipe and esophagus when a goat swallows

Windpipe and esophagus when a goat breathes

1. Before inserting the tube, lay it along the kid's side with the rounded tip extending to the last rib and the other end at its lips. Mark that top point so you will know how far to safely insert the tube.

2. Slowly and gently push the tube down the kid's throat. Often, the kid will swallow the tube as you advance it, which helps.

3. When the end of the tube reaches the stomach, attach the syringe of colostrum or milk — with the plunger removed — to the upper end of the tube.

4. Allow the liquid to naturally flow directly into the stomach. Wait a few extra seconds to be sure the tube is empty before slowly removing the tube from the throat.

Note: Be absolutely certain that the tube is in the stomach before administering nourishment, so the liquid isn't forced into the lungs. It's a little tricky to know for sure with kids that haven't had any milk, although they tend to cough if the tube is in the lungs. You can estimate the length of tube needed to reach the kid's stomach to get a fair idea of how far it will need to be inserted. If the tube is stopped before you think it should, you are probably in the wrong place and should start again. If you need to tube an older kid at some time, blow gently into the inserted tube. A properly placed tube will make a gurgling sound. Air coming out of the tube will also have a distinctive curdled milk smell.

Once a kid is cleaned off and on its feet, it will make its way to the doe's udder for its first drink of colostrum. The job is made easier if the doe gets a dairy clip before she kids, as this new mother has.

house until the weather warms up. Even then you should harden it off by degrees rather than exposing it to the cold all at once.

Colostrum

The kids will be standing and trying to walk on wobbly legs, perhaps within minutes of birth, and they'll soon be looking for their first meal. This must be their dam's colostrum, or "first milk," a thick, sticky, yellow, nonfoaming milk. Colostrum contains important antibodies and vitamins, and the survival of any newborn mammal is in jeopardy without it. It's so important, in fact, that you might have health problems with kids born to new does in your

RULE OF THUMB

Ensure that each newborn kid will receive 1 ounce (30 cc) of colostrum per pound (0.5 kg) of body weight, within the first 12 hours and about that much in the next 12. The sooner the colostrum is fed, the better. Studies show that the ability of the intestinal wall of the newborn to absorb antibodies quickly decreases. By 18 hours after birth, there is very little absorption, and by 24 hours there is none. That is nature's way of protecting the kid from foreign infection.

barn. If she has not had time to build up immunities to your barn's bacteria, they can't pass to her kid through the colostrum. Antibodies provided by the doe's colostrum give the kid immune support for 3 to 6 weeks, depending on the disease. After that time, boosters for things like tetanus and enterotoxemia should be considered (see chapter 8).

You can allow kids to nurse, or you can milk out the doe and feed the kids from a bottle. For meat kids, nursing is fine, but for doelings that you want to sell or keep as future milkers, milking the mother is preferred. It is not only to prevent CAE and Johne's disease and because hand-fed kids are easier to handle as adults, but also because it's the only way you can be certain that the doe is producing colostrum and the kids are getting enough — and not too much.

Either way, it's extremely important to get some colostrum into them within half an hour or so of birth. Put it into a clean soda bottle for feeding and use the nipple that you will use to feed the kid as it grows. There are special valve nipples that are good for very tiny or premature babies, but a standard caprine nipple that can later be fitted into a gang feeder (also called a "lamb bar") works best. At one time, pan feeding was popular, but experienced goat owners have realized that it is too easy for kids, flies, cats, and bedding to contaminate milk in a pan or trough, and there is no way to be sure which kid is getting how much. Pan feeding also is a common cause of bloat (see page 126). A normal first feeding is about 8 ounces (235 cc). *Warning:* Never heat colostrum in a microwave, and never heat it above 140°F (60°C) or you'll destroy the antibodies. If it must be warmed back up to goat temperature, set the bottle in warm water for several minutes (see page 177 for instructions on heat treating).

If for some reason the doe has no colostrum, you can make an emergency substitute.

Colostrum Substitute

- 3 cups milk
- 1 beaten egg
- 1 teaspoon cod liver oil
- 1 tablespoon sugar

Mix well. This, of course, lacks the maternal antibodies that provide immunity from common diseases of the newborn, but it will provide the laxative function of colostrum and help clear the kid's intestines of the tarry, black substance called meconium.

COLOSTRUM FOR HUMANS?

Most goat owners feed all the colostrum their does produce to the kids, or they freeze any extra for emergencies, such as when kids are orphaned or a doe doesn't produce enough. But if enough is available, it can be used as human food.

Called "beestings" in England, 1 cup of colostrum can replace 2 eggs and a scant cup of milk in baked goods. A custard can be made by mixing 2 cups of colostrum with 2 cups of milk and about ¼ cup of honey. Bake at 300°F (149°C) until set, 45 to 60 minutes. Sprinkle with nutmeg if desired.

Commercial colostrums are also available, but a study from the Netherlands released in 2008 said if goat colostrum is not available, cow colostrum is the best alternative. Artificial colostrum was the least able to provide adequate levels of protection for the kids. Even cow colostrum can cause some problems because of the dangers of Johne's disease that can be transmitted from cows to goats. Goat colostrum from another farm has its drawbacks too, because those goats won't have immunities to the same bacteria that grow on your farm where the kids will be raised. Be aware, then, that there is no good substitute for the real thing from the real mom.

If you are on a CAE prevention program, colostrum must be heat-treated. Although commonly referred to as pasteurization, that's not technically correct, since temperatures required for pasteurization would turn colostrum into a pudding. Colostrum must be heated (preferably in a double boiler) to at least 135°F (57°C) but no more than 140°F (60°C) and held at that temperature for 1 full hour (see box at right.) Since it takes at least an hour to heat-treat colostrum, and kids should be fed within a half hour of birth, it is wise to save extra treated colostrum in the freezer so it is ready for subsequent babies. Pour it into clean plastic soda bottles and keep it in the freezer. Defrost it in warm water, pop a nipple on the bottle, and feed. Bottles can be kept until the next year but not in a freezer with an automatic defrost cycle.

HEAT-TREATING COLOSTRUM

Heat-treating colostrum and milk for kids has become common since CAE became a problem. But note that these two forms of milk require two different procedures.

Milk is pasteurized by heating to 145°F (63°C) and maintaining that temperature for 30 minutes. Alternatively, it can be heated to 161°F (72°C) for at least 15 seconds. This is the method used in processing plants, but it's more trouble for the home dairy.

Colostrum, however, is altogether different. If it gets too hot, it becomes pudding. To heat-treat colostrum, warm it to at least 135°F (57°C) but no more than 140°F (60°C) and keep it at that temperature for 1 full hour. One simple way to accomplish this is to bring it up to the proper temperature in a double boiler or water bath (to avoid scorching). Then pour it into a preheated thermos, and let it stand for an hour. Be sure you have an accurate thermometer and a quality thermos that will actually maintain that temperature for that period.

Cool the colostrum to about 110°F (43°C) before feeding. (The ideal feeding temperature is about 103°F [39°C], but it will cool to that by the time the kid starts drinking.)

12

RAISING KIDS

With the excitement of freshening over, your goat barn can settle into a routine. For the first 3 or 4 days your doe will produce colostrum, the thick yellow milk so necessary for the kid's well-being. If you used an antibiotic treatment when you dried off the doe, wait at least 7 days before using her milk for human consumption. That is twice as long as the withholding time on the label for cows, but goat owners who regularly use dry treatments and have their milk tested for antibiotic residue find that the antibiotic has cleared the system by day 7.

By then, there should also be enough milk for both the kids and you. (After perhaps months of anticipation, what a treat!) And after 2 months or more of relative inactivity, your goat barn will be a hectic place. In addition to the usual feeding and cleaning tasks, you'll be milking twice a day — and raising kids.

Raising kids requires some knowledge and a lot of work and time. Goat raisers have many different opinions about how the job should be done, but none can deny that the first year of the goat's life, along with her breeding and prenatal care, is an important determinant in how she will behave and produce later. If you are also raising kids for meat, much of the same information applies, with the possible exception that a meat kid might as well nurse its mother as long as possible and save you the trouble of bottles and separate pens. If the kid swipes the milk that you had planned for supper, you might need to devise a plan to keep Mom on her own until milking time and then let the kids in when you have enough for your family.

Early Feeding

If the doe has a congested udder or a very hard udder, the condition often can be helped by letting the kids nurse for the first few days — unless, of course, you are on a CAE prevention program. The suggested procedure is to bring the kids to their dam every few hours, rather than leaving them together. While this entails more work, it eliminates a lot of commotion and consternation later on when you expect them to drink out of a bottle or a group feeder. First fresheners, which often have very small teats, are also frequently left with their kids if the milker's hands are too large. The teats will enlarge with time.

If you milk the doe, do it within half an hour of kidding and offer the kids some colostrum. It should be close to goat body temperature, which averages around 103°F (40°C). Colostrum scorches easily: use a water bath or double boiler to warm it (see chapter 11).

The antics of young kids frolicking are so entertaining that most people can watch them for hours!

How will the kids be fed? Nursing is certainly the easiest method but not necessarily the best, as far as goat breeders are concerned. Some people say it ruins the dam's udder, which is important not only if you intend to show her but also if you want her to have a long and productive life as a milker (of course, others say that's a lot of bunk and are quick to show examples of does in their herds that are on their eighth set of kids and have lovely udders).

Possibly a more important consideration is that you don't know how much the doe is producing or how much the kids are getting. Also, kids left with their mothers are much wilder than hand-raised kids. Another important consideration is that once a kid learns to suck its dam it will be difficult — maybe impossible — to teach it not to. Some does wean their kids relatively early, but there have been other cases where yearlings are still sucking, and your milk supply for the kitchen is lost. The only solution in those cases is complete separation. It's better to do it right away.

If you're concerned about caprine arthritis encephalitis or Johne's disease, you don't want the kids to nurse at all (see page 130).

Pan Feeding or Bottle Feeding?

Kids not left with their dams can be pan fed or bottle fed.

Many breeders prefer bottle feeding because it's more "natural." They point out that with pan feeding the animal is forced to lower its head to drink and milk can get into the rumen where it doesn't belong. Digestive upsets can result. Of more immediate practical concern are the serious problems of fecal contamination, spilling (think of it as a sour milk fly attractant), and skin irritation where

A KID IN HAND . . .

It's a good idea to handle your kid as soon as possible and get her used to having your hands all over her — especially her belly, legs, and where her udder will eventually develop. Life will be much nicer when it's time to put her on a milk stand as a yearling.

ears and feet have flopped in the milk. Still, pans are much easier to fill, wash, and sterilize than bottles and nipples; you won't need lamb nipples or bottle brushes. A milk trough made of a PVC pipe with caps on each end can be set up on blocks to eliminate some of the problems of pan feeding.

It isn't necessary to have a bottle for each kid. A variation that's very popular where large numbers of kids are fed is a large container (such as a 5-gallon [20 L] pail) with special nipples. The nipples are attached to plastic tubes that reach to the bottom of the container. The kids suck on the nipples, and the milk is drawn up through the tubes, just like drinking through a straw. Commercial units go by such names as Lambar and Lamb Saver. You can purchase the nipples and tubes and make one from a 5-gallon bucket that will feed up to 10 kids at a time. The nipples also fit over soda or beer bottles so you can start the kids on the nipple, and they will easily shift from a single bottle to the gang feeder. By putting the straws into individual quart jars of milk fitted down into the bucket, you can also tell how much any one kid is drinking.

Several brands of nipple buckets like this are available commercially, or you can make your own. This model works on gravity, but others have a special nipple attached to plastic tubes that kids quickly learn to suck like a soda straw.

MAKE YOUR OWN BOTTLE RACK

In a 1 × 4 of any length you want, cut or drill holes large enough to admit the necks of the bottles you'll be using. Nail the 1 × 4 at a right angle to a 1 × 10 or 1 × 12 to form an L; this allows the bottles to rest on the larger board with just their necks and the nipples poking through the holes. Using a 1 × 6 board for backing, fasten the rack to a fence at about a 45-degree angle and at a convenient height, depending on the size of the kids. Finish sides with a piece of ½-inch plywood as shown below.

Hungry kids butt udders (and bottles), and when they get large enough, you'll probably need another device to hold the bottles more firmly in place. And occasionally, a kid will pull a nipple off a bottle, dumping the milk. But overall, we've found this rack to save a great deal of time and labor.

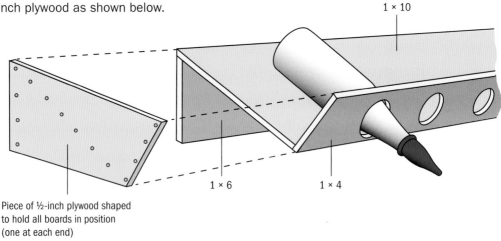

1 × 10

1 × 6   1 × 4

Piece of ½-inch plywood shaped
to hold all boards in position
(one at each end)

If you're feeding more than a few kids with bottles, bottle racks are handy. You can feed as many kids as you want to at one time, without holding the bottles in your hands. You can buy plastic bottles with lamb nipples and wire racks that attach to board fencing, or you can easily construct a rack to hold soda or beer bottles (see page 183). Glass bottles are notorious for building up a vacuum and frustrating kids as they try to suck. One trick is to lay a single strand of rubber band across the bottle opening, then slide the nipple on. The part of the rubber band hanging down the sides will allow a little air into the bottle as the kid sucks.

Whichever feeding method you decide to use—pan, bottle, or nursing—once you start it, you're stuck with it. It's difficult to teach a kid to drink from a bottle once it's used to a pan, and the other way around, although it is possible to go from dam to bottle with a little patience.

Likewise, after the first feedings of colostrum, milk may be fed warm or cool, but be consistent to avoid digestive upsets. Feeding on a regular schedule is important for the same reason. Kids will consume less milk if it's cold. Most people feed it lukewarm.

Frequency

Everyone who has raised a few kids seems to have an opinion about the best method, in terms of nursing, pans, bottles, frequency, amounts, and weaning time. The fact that their kids survive and thrive is evidence that all of these methods work — in any combination. What it comes down to, then, is a matter of personal preference and convenience.

As with any baby, frequent small feedings are better than infrequent large feedings. Some people feed their kids every few hours. But if you have a job or a busy schedule, this might not be possible. In that case, feeding every 12 hours will be more convenient. Either method works, as long as you're consistent.

Of course, there are limits and guidelines. The kid should get its first colostrum feeding as soon as possible after it is born and its second within the first 12 hours. The goal is 1 ounce (29.6 cc) per pound (0.45 kg) of body weight in the first 12 hours. They generally are hungry at the first feeding and may take as much as 8 ounces (236.6 cc) but may not be as hungry at the second. The third feeding should be within the next 12 hours. Colostrum is critical to survivability, so be conscientious about these first feedings. Over the next 6 days, the recommendation is 3 feedings a day of milk, milk replacer, or a mixture of the two starting at about 6 ounces (177.4 cc) per feeding and working up to about 10 ounces (295.7 cc). During their second week, they will probably

work up to 12 ounces (354.9 cc), at each of three feedings 8 hours apart. And from then on you can expect to feed about 1 quart (1 L) of milk to each kid, twice a day.

But none of this is rigid or a matter of life or death. Some kids simply won't want this much. If they don't act sickly, they're okay. Some people give the kids all they will drink, and that's okay too, as long as the kids don't scour (if they do, see the advice on scours in chapter 8, which includes limiting milk intake).

If you feel lost without specific directions to follow, these will help. But remember that observing your animals is the real key to good husbandry! This approach is far better than saying a kid must consume a certain amount each day. If they act hungry, check the amount you're providing and their body weight to make sure you're feeding them enough. If they don't want all you offer, check to make sure they're just full, not sick. If the kid's stomach feels hard when it has finished eating, it is overfull and runs the risk of bloat. It is best to leave it a little underfed.

As a rule of thumb, expect a newborn kid to consume about 2 cups (0.5 L) of colostrum a day, in three or four evenly spaced feedings. This gradually increases to roughly 2 or 3 pints (1 or 1.5 L) of milk or milk replacer a day at weaning, if they're weaned at 8 to 10 weeks. Offer warm water after each feeding of milk.

It's best to leave them a little on the hungry side, as this will encourage the consumption of solid foods, which helps develop the rumen. They will start nibbling on fine-stemmed hay in a week or so and on grain (18 percent kid starter) soon after. The more solids they consume, the less milk they'll drink.

One way to judge your relative success is to weigh each kid once a month. Ideally, they should gain about 10 pounds per month, for the first 5 months. For example, a kid weighing 8 pounds at birth would then weigh about 38 pounds at 3 months of age.

Weaning is another controversial area. On average, it's a safe guess to say that most kids are completely off milk by 8 to 10 weeks of age.

Still, some people feed milk for much longer, as long as 6 months, in some cases.

MODERATION IS KEY

This is one area where the old saying, "The eye of the master fatteneth the cattle," is especially true. You obviously don't want to starve the kids, but don't kill them with kindness either. This happens most frequently by overfeeding milk and causing scours, which can be fatal. You don't want kids to be "fat and healthy" because fat isn't healthy for a dairy animal. Strive for condition, not overcondition. The kid should be producing bone, not fat, to develop her full potential in later life.

SUGGESTED FEEDING SCHEDULE FOR KIDS

Age	Feed	Amount	Frequency
First 12 hours	Colostrum	1 oz/pound of weight	2 times
Next 12 hours	Colostrum	1 oz/pound of weight	1 time
2–7 days	Milk, replacer or mixed	6–10 oz	2–3 times a day
1 week	Milk, replacer or mixed	10–12 oz	2–3 times a day
2 weeks	Milk, replacer or mixed	12 oz	2–3 times a day
	Good hay	Free choice	
	Water	Free choice	
3–8 weeks	Milk, replacer or mixed	12–16 oz	2 times a day
	Good hay	Free choice	
	18% starter grain	As much as a kid will clean up in about 15 minutes	2 times a day
	Water	Free choice	

Note: These amounts are approximate and are guidelines only. Base actual amounts on the appetite and condition of your kids. Coccidiosis is a protozoal disease of kids between 3 weeks and 5 months old. Kids can become infected even on the most scrupulously managed farm. Because of that, you may want to consider using feed or water supplements that contain a preventive level of coccidiostat.

Some goat owners prefer to feed kids individually so they know exactly how much each kid is eating. The Pritchard nipples on these bottles are made especially for small babies and should be exchanged for a larger nipple as these little meat kids grow.

This certainly isn't necessary, and according to Dr. Lennart Krook of Cornell University, it may actually be harmful. Kids overfed calcium (milk is high in calcium) are likely to develop bone troubles in later life. In addition, we want that rumen to develop. That requires hay.

Milk Replacer

If you want to keep all the milk your goat produces for yourself, the kids can be fed milk replacer. However, be sure to use milk replacer made for sheep or goats, not cows. Calf milk replacer is not high enough in fat, and goats will not do well on it. In fact, a large commercial dairy that has tried several brands of milk replacers and keeps meticulous records on all its goats claims that, even when the kids looked fine while on cow milk replacer, 2 years later most of them were dead or had been culled because they lacked stamina.

Milk replacer should be 16 to 24 percent fat and 20 to 28 percent protein with milk-based proteins. A good option is to mix milk replacer with whatever goat milk you don't need for the kitchen.

Milk replacer is more likely to produce bloat in kids than is goat milk, especially if the kid eats too much. After mixing, let the milk settle down so the kid isn't drinking extra air bubbles that can upset its stomach. Other than these few tips, there are very few hard-and-fast rules about anything connected with kid raising.

Weaning

Early development of the rumen is extremely important for later production. Most kids will start to nibble at fine hay by the time they're 1 week old. They should be encouraged to do so with kid-size mangers and frequent feedings of fresh, leafy hay. Hay or forage is more important than grain.

This also translates into limit-feeding milk. When you limit milk consumption to 2 pints (1 L) a day, you encourage consumption of dry feed. This increases body capacity, with a corresponding increase in feed intake and digestion. Research has shown that at 2 months of age a weaned kid has a reticuloruminal capacity five times larger than a suckling kid of the same age.

Within a few weeks of birth, kids will start testing solid food. They won't gain any nutritional value from the forage at first, but this is the first step in the process of their stomachs shifting from metabolizing milk to digesting forage.

Wean by weight, not by age. The usual goal is two to two-and-a-half times the birth weight. The primary consideration should be whether the kid is consuming enough forage and concentrate to continue to grow and develop without milk.

At weaning, most breeders feed a commercial kid starter or calf ration with a coccidiostat: ½ pound (227 g), twice a day. The kids should always have access to good hay. At 6 months they are switched to a milking ration. By 7 months, doelings weigh 75 to 80 pounds (34 to 36 kg) and are bred. Milk-fed kids weighing 20 to 30 pounds (9 to 14 kg) are in great demand as meat in some localities, especially at Easter and Passover.

Castration

If bucks will be slaughtered at a few months of age, castration is of little, or even negative, value. Intact males grow faster because of a full flush of muscle-bulking hormones and the absence of the stress caused by castration. If they will be held longer, they should be castrated before weaning to avoid a "buck odor" in the meat. The exception to the rule is when you are marketing to an ethnic group that demands intact kids. Check with the local auction barn or meat buyer before you make your decision, and be sure to segregate those bucklings from your doe herd. Buck kids kept for meat require no special diet, but some chevon (goat meat) aficionados claim milk and browse produce the best meat. Butcher kids don't need a lot of grain. It has little effect on the palatability of the meat because the fat is deposited in the kidney and pelvic regions rather than in the muscle, and too much grain can lead to urinary calculi and other problems.

No matter which method of castration is used, a tetanus vaccine is required. This is usually given at 3 to 4 weeks of age, along with a vaccination for *Clostridium perfringens* types C and D, the bacteria that cause enterotoxemia.

Until they are about 1 month old, bucks can be surgically castrated without anesthesia. They can also be neutered with a small Burdizzo emasculator or an elastrator. After about 1 month, the surgical procedure should be used and with an anesthetic. A chemical method is also available.

Surgical Castration

A sharp, sterilized scalpel and an assistant are needed for this procedure.

WEAN BY WEIGHT

Kids can be weaned by age or by weight. While it's possible to wean kids as young as 4 weeks old, 8 weeks is considered optimal. Later weaning costs more in milk and labor, and it retards rumen development. But weaning by weight is better because it prevents unhealthy or undernourished kids from being weaned too early and can reduce weaning stress. Studies suggest that weaning at a total body weight that is two-and-a-half times the birth weight produces good results. This is usually 19 to 22 pounds (8.5 to 10 kg).

A helper holds the kid by the hind legs, its back to the helper's chest. A quick, clean incision through the scrotum is made with the knife, and the testicle is grasped and pulled out. The other testicle is likewise removed and the wounds sprayed with antiseptic.

Since this is a surgical procedure, it's best left to a trained veterinarian. In fact, in England it's the law that goats over 2 months old can be castrated only by veterinary practitioners, using anesthesia. Obviously, this is nothing to be taken lightly by an amateur who lacks knowledge, training, and experience.

Using an Elastrator

It's also possible to castrate with strong, tight rubber bands made for the purpose and applied with a special tool. Both are available at most farm-supply stores. The special bands are slipped over the scrotum above the testes. Once the bands are applied, the testicles atrophy and the bands and sac fall off in a few weeks. Some people feel the practice of castrating this way is inhumane, because the buckling is in obvious discomfort for about a half hour after the band is secured, and there is a danger of tetanus if the kid has not been vaccinated.

BURDIZZO

The Burdizzo, or emasculator, is used to crush the testicle cords to neuter young males. These are available from farm-supply stores and catalogs.

On the other hand, the equipment is inexpensive and readily available, the job can be done by anyone who can count to two (two testicles below the band), and the buckling forgets the band is in place within an hour. There is no blood or the potential fly aggravation that comes with surgical castrations, and there are no accidental kids that come from missing the mark when using a Burdizzo. For the farmstead goat raiser, there are many choices to be made, and the method of castration is just one of them.

FINAL THOUGHTS

Disbudding and tattooing are also important steps in getting your kids off to a good start. See chapter 7 for more information.

ELASTRATOR

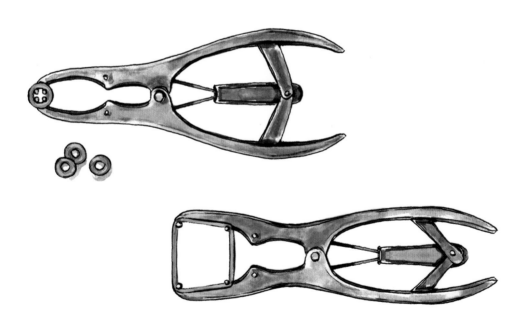

The elastrator uses special, small, very strong rubber bands to neuter buck kids. This tool can also be used for dehorning, although this method may not be the best choice (see page 115).

13

MILKING

The new dairy goat raiser must learn about goat feeds and nutrition, about bucks and breeding and raising kids, all for one purpose: to get milk. Milking, therefore, is at the apex of the pyramid of all goat-keeping skills.

Most goats are generally milked at regular 12-hour intervals and according to a regular routine. Milking at 6 a.m. one day and 9 a.m. the next is one of the easiest ways to depress milk production. You might milk at 7 a.m. and 7 p.m. or at noon and midnight, but it should always be as close to 12 hours apart as possible and always at the same time (see appendix, page 252 for an explanation of this).

This has been the conventional wisdom, which most people follow. However, recent studies in France indicate that milk production in goats isn't affected if other schedules are followed, such as milking 10 and 14 hours apart, or even 8 and 16 hours apart, as long as the same schedule is followed every day. Whatever your schedule, once you become a goat milker, your daily routine will be set not by your favorite television show but by the goats.

Of even more importance than regularity is sanitation. One of the main reasons for keeping goats is having milk better than any to be found in the supermarket dairy case. This requires not only a knowledge of dairy sanitation but also a rigid adherence to sanitation principles.

Milking Essentials

What do you need to make milking comfortable and efficient? Equipment and good sanitation.

Equipment

Milking equipment can be simple or elaborate. You could, if you wanted to, milk into a bowl from the kitchen cupboard, make a milk strainer from two inexpensive funnels, and store the milk in the refrigerator in fruit jars. At the opposite extreme you could spend hundreds or even thousands of dollars on milking equipment.

If you intend to milk 730 times a year for the foreseeable future, you will get more satisfaction and better quality from proper equipment.

Milking Pail

A 4-quart (3.75 L) seamless, stainless steel, half-moon hooded milking pail is no luxury, despite its fairly high price. Many of these have been in daily use for 20 years or more, which brings the per-use cost down to a pittance. Being seamless

A milking stand like this makes that pleasant twice-a-day chore even easier and more enjoyable. Goats quickly learn to jump onto the stand at milking time.

and stainless, these pails are easy to clean and disinfect. The hood (which keeps hair and dust out of the milk) and handle are removable to enable thorough cleaning. They're made especially for goats, naturally, so you'll have to check the goat-supply houses currently advertising in goat publications to find one. This is the one piece of goat equipment I couldn't do without.

Whatever you use, avoid plastic. No amount of cleaning can get the bacteria out of the pores in plastic, and you'll soon end up with a product that's unfit for human consumption.

Some people refuse to use any dairy equipment made of aluminum. This might be due to some misconceptions. It's true that aluminum isn't allowed in grade A dairies, but that's because their equipment must endure a great deal of very vigorous cleaning. The aluminum gets scratches, resulting in the problems associated with plastic. Some are no doubt concerned about reported connections between aluminum cookware and human health, especially Alzheimer's disease. But you don't cook milk in your milk pail or strainer.

Aluminum home-dairy equipment is anodized, or coated with a protective film. Milk never touches the aluminum unless the coating is removed by abrasive cleaning. The nicks and scratches that lead to potential problems in the commercial dairy can be avoided in small operations with proper care and procedures (see pages 206–207).

Strainer

A strainer is a necessity. It must be of a type that has a removable spring clamp in the bottom that holds disposable milk strainer pads, available at any farm-supply store. The bottom of the strainer must be small enough to fit into your holding container, such as a widemouthed canning jar. Small 1-quart (1 L) tin kitchen strainers are inexpensive and work well with small amounts of milk, but larger half- and one-gallon (2 to 3.75 L) sizes made especially for backyard dairies are available in both stainless steel and aluminum (stainless steel costs about twice as much but doesn't scratch and will be around for years).

Milk filters are used only once and then discarded. Reusing a filter or running milk through a recycled cheesecloth are both surefire recipes for high bacteria counts and really bad milk quality.

If you're on a very tight budget, you can make a strainer from two large kitchen funnels. Cut off the spouts, and a little more, so you have two funnels with openings of 2 to 3 inches (5 to 7.5 cm). Put a milk filter pad into one funnel, and place the other funnel on top to hold it down firmly.

Milk can be stored in 1-, 2-, or 4-quart (1, 2, or 3.75 L) glass jars or in aluminum or stainless steel cans of the same size. Again, avoid plastic. Look for something easy to clean and sterilize.

Udder Washing Supplies

In addition to these tools, the home goat dairy will require a bucket to hold udder wash or, better, a spray bottle that will eliminate passing contamination from one goat to the next. Generally, these are used only if the udder is particularly dirty at milking time. You can use an udder wash, but the recommended practice of predipping with a teat dip will require commercial teat dip and a dip cup, disposable paper towels for drying the udder, udder wash, and cleansers and disinfectants for utensils. Commercial dairy owners use plastic gloves to prevent the spread of disease from one goat to another. The process of dipping and wiping udders keeps them fairly clean, and they can be easily rinsed off if they get too messy. In a family dairy with only a few goats, scrupulously clean hands are probably just as effective and much less expensive.

Scale

A scale for weighing milk so you can record production isn't an absolute necessity, but it's a very good idea. Scales usually come with two adjustable hands, but you don't need to use them both. Hang the empty milk pail on the scale hook and set the main hand at "zero." When the bucket is full of milk and hung back on the scale, the hand will be pointing to the weight of the milk and not the added weight of the pail.

Strip Cup

A strip cup, used to detect abnormal milk and mastitis, is simply a metal cup with either a screen or a black tray at the top. You squirt

Milking equipment includes a hooded stainless steel milk pail, strip cup, milk strainer with filter disks, milk can (or tote pail), teat dip cup, and hanging scale. Milk pails come in 2- and 4-quart (2 and 3.75 L) sizes and can be stainless steel or aluminum, but glass containers can also be used. Avoid plastic.

HOW TO MAKE A FOLDING MILKING STAND

MATERIALS

For platform

- Two 1 × 8 × 42" floorboards
- Two 1 × 4 × 15" cleats
- One 1 × 8 × 32" seat
- One 1 × 4 × 15" leg for platform
- One 1 × 4 × 14" leg for seat

For stanchion

- Two 1 × 6 × 46" board (stanchion)
- Two 1 × 4 × 14" top cleats
- Two 1 × 4 × 14" bottom cleats
- One ¾ × 14 × 14" plywood for feed-pan holder

HARDWARE

- Screws (or nails) (1¼" and 2¼")
- Two 4" T hinges for legs
- Two 4" strap hinges for stanchion
- Two 3" strap hinges for feed-pan holder
- Two 5" heavy strap hinges for platform
- 1 eye hook with 2 screw eyes for stanchion and stand
- 1 eye hook for feed-pan holder

We've been using a folding, wall-mounted milking stand like this one for more than 40 years and highly recommend it. It's more comfortable to use than the common bench style (without the seat), and the folding feature makes it a real space saver. It can be built in a couple of hours for less than $60.

If, like most homesteaders, you have a "treasure pile" of recycled lumber on hand, the only expense would be the hardware. Unlike most milking stands, this one consists of two parts: a platform for the goat with a seat for the milker and a stanchion to restrain the goat's head and to hold a feed pan.

Important Note: The nominal size of the stand is 42 inches (1 m) long by 15 inches (0.3 m) wide, but the size can be adjusted to fit your goats or the materials you might have on hand. Also, recall that modern lumber dimensions don't match the names. For example, a 1 × 6 board is actually ¾" × 5½". Such details aren't critical for this project.

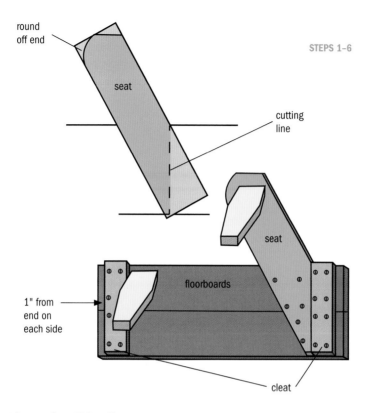

round
off end

STEPS 1–6

seat

cutting
line

seat

floorboards

1" from
end on
each side

cleat

Constructing the Platform

1. Lay the two 1 × 8 × 42" floorboards side by side (rough side down, if you're using rough lumber that has one side better than the other).

2. Place a 1 × 4 × 15" cleat 1" from each end of the floorboards, as shown.

3. Fasten the cleats to the floorboards with screws or nails, making sure they don't extend through the floor. If nails protrude, bend them over and clinch them well.

4. Lay the 1 × 8 × 32" seat board on top of the platform so one corner protrudes beyond the platform by about an inch and the other side is flush with the cleat (see drawing). The angle of the seat isn't of extreme importance, but this method yields a good angle for most milkers.

5. Mark the seat with a cutting line along the cleat as shown. Cut the seat board so it fits snugly against the cleat. Trim off the other scrap.

6. Round off the end of the seat, and secure it to the platform as you did the cleats.

Instructions continued on next page

Constructing the Platform (continued)

7. Taper the legs as shown, if you wish, or leave them square. Note that one leg is longer than the other.

8. Fasten the 1 × 4 × 15" leg to the underside of the platform, next to the rear cleat, with a 4" T hinge.

9. Fasten the 1 × 4 × 14" leg near the end of the seat with the other T hinge, as shown.

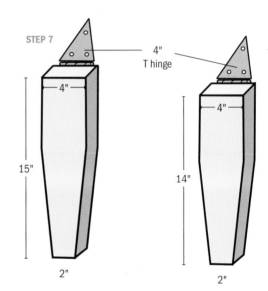

STEP 7

4"
T hinge

4"

4"

15"

14"

2"

2"

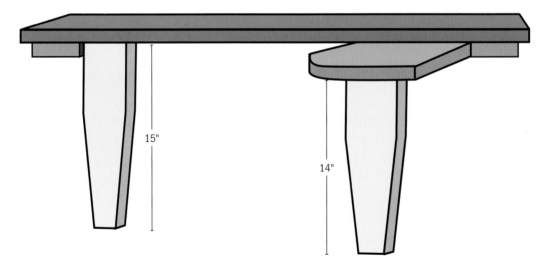

STEPS 8 AND 9

15"

14"

Constructing the Stanchion

1. Lay the two 1 × 6 × 46" boards about 4" apart, or so they line up with the width of the platform, minus 1" (if you have changed the platform dimensions, check this carefully, or the screw eye that holds the stanchion to the platform won't line up properly. The 1" difference allows the stanchion to fold over the platform). For most goats a neck space in the stanchion of 3½" to 4" (9 to 10 cm) is good.

2. Screw or nail two 1 × 4 × 14" cleats across the top and two across the bottom of the stanchion (you need two cleats for each end so the stanchion can fold over the platform when it's not open for use).

3. Center and draw an 8" circle just below the top cleat. Cut out as shown.

4. Center, draw, and cut out a circle in the ¾ × 14 × 14" board that will hold the feed pan. The size of the feed pan you intend to use will determine the size of the circle.

5. Attach the two 4" strap hinges to the top and bottom cleats, as shown.

6. Using two 3" strap hinges, hinge the feed-pan holder to the other side, about 22" from the bottom of the stanchion. Attach hinges to the underside of the feed-pan holder.

4" strap hinge

8" circle

3" strap hinge

22"

hole for feed pan

cleat

Instructions continued on next page

Mounting the Stand

1. You'll mount the platform to the wall first. Determine where the wall's studs are, and space the hinges accordingly. Attach the two 5" strap hinges to the platform.

2. Ensure that the platform is level before attaching it to the wall. Rather than measuring the distance from the floor to the hinges or from the floor to the platform, get the platform reasonably level before marking where the hinges will be attached to the wall, as the floor might not be level. If the floor slopes away from the wall, for example, the hinges might need to be less than 15" from the floor.

3. Secure the platform's hinges to the wall's studs.

4. Mount the stanchion to the wall. Ensure that the top of the stanchion's bottom cleat is snug with the underside of the platform for support. Also, be sure to leave a 1" space between the stanchion and the wall (you need this clearance in order to close the stanchion over the platform when the unit is folded; this is why the stanchion is 1" narrower than the platform). Screw the stanchion's 4" strap hinges to the wall, ideally through a stud.

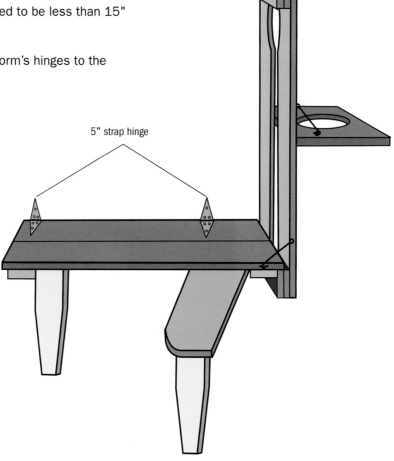

5" strap hinge

5. Check that the platform and stanchion open and close properly and are reasonably plumb, square, and level.

6. Install an eye hook to hold the platform and stanchion together when the stand is open for business. Then fold the stand and, using the same hook, install another screw eye to hold the stand folded.

7. Install an eye hook and screw eye to hold the feed-pan holder level.

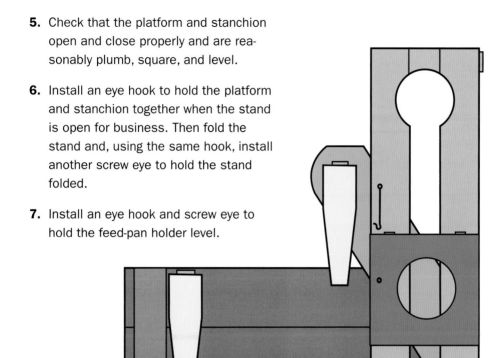

MOUNTING TIP

Check the stud spacing in the wall before mounting the milking platform to it. Screwing the hinges to studs provides a more secure installation. You don't want to attach the stand to drywall or plywood, for example.

Milking Time

Now you're ready to milk. Once your goats discover that there is feed in the pan, they'll readily jump up on the stand at milking time. When your work is done, fold the stand out of the way.

the first stream of milk from each teat into the cup and examine it for flakes, lumps, and other signs of abnormality. A strip cup will help you maintain a constant check on one aspect of your herd's health and the quality of the milk your family drinks.

Milking Stand

A milking stand is far more comfortable than squatting, especially if you have a number of milking does or if you tend to creak anyway. Milking stands can have stanchions to lock the doe's head in place to help control her while you're milking, and a rack to hold a feed pan to keep her occupied. A doe will quickly learn to jump up on the stand at milking time, especially if she knows there's grain waiting there.

Facilities: A Milking Parlor or In-Barn Milking

Ideally, the milking should be done in a special room, away from the goat pen and any sources of dust and odor. It should have good ventilation, running water, electricity, a drain, a minimum of shelves or other flat surfaces that gather dust, and impervious floor, walls, and ceiling so it can be kept even cleaner than your kitchen. All this, and more, is required for licensed commercial milk producers.

Alas, ideals are often unattainable for us poor goatherds. Many people with just a few goats milk in the aisle of the barn but not in the pen.

Preparing to Milk

Whether milked in a separate parlor or in the goat barn, the goat should be brushed to remove loose hair and dust before milking. If her udder is particularly dirty — which is rare for goats — she should be washed with a dairy disinfectant (made for the animals, not the utensils) and thoroughly dried. A hooded pail with a removable carrying lid should be used to collect the milk. It's also very important that the milker have clean, dry hands. Wash and dry them before milking each doe.

Enjoy your milking time, wherever you do it, but remember to protect the milk from contamination if you will be using it for food.

DUST, BACTERIA, AND HAIR

Goat milk actually has fewer bacteria than cow milk as it leaves the udder. But on the debit side the goat milk is more likely to pick up coliform bacteria during the milking process. This is due in part to the dry nature of goat dung; it actually becomes dusty. Combined with the loose housing most commonly used for goats, this results in dung dust and coliform bacteria in the air, on flat surfaces, and on the goat. And the goat milker is more likely to disturb the hair on the belly than the cow milker just because of the size of the animal. There might also be more of it, so hairy goats should be trimmed, even if it's only a "dairy clip" around the udder area during cold weather.

Milking Procedure

Finally, we're ready for the actual milking. We'll take it step by step.

It looks easy, until you try it. But then with a little practice it is easy, and you'll wonder why you had milk up your sleeves and all over the wall and your legs the first time you tried.

Get into Position

Position yourself at the goat's side, facing the rear. Goats can be milked from either side, but they develop a definite preference for the side they're milked from. Many commercial dairymen milk from the rear with machines, but it puts your face and nose in a vulnerable position when milking by hand, so it's not recommended.

Predip

Until recent years, the recommended process was to wash the udder with warm water and an udder-washing solution and then dry with a fresh paper towel for each goat. Studies have found that the extra water is a very effective carrier of bacteria, and the place it collects is right at the orifice of the teat where it can do the most harm. A properly mixed commercial udder wash offers some antibacterial control, but it is getting hard to find in light of the gradual shift to predipping with a teat dip. It's still a good idea to use an udder wash for prepping dirty or muddy udders, but the more effective mastitis-fighting predip is recommended. Use a 0.5 percent iodine predip. Bacteria levels have been shown to be five to six times higher in milk acquired without predipping than in milk acquired with a predip. The risk of *Listeria* is four times higher without a predip.

Fill a small kitchen cup or a teat cup designed for the purpose with an iodine-based commercial teat dip. (In time, you can experiment with a teat dip that fits your needs. Some of the blue-tinted dips may be too harsh for the goat's tender skin and should be stopped if any irritation is apparent.) Dip at least three-quarters of each teat in the dip. It must stay on the teat at least 30 seconds to kill bacteria effectively. "Wash" the dampened teat with your hand. Many people prefer wearing disposable latex or nitrile gloves, which are easy to clean and less likely than bare hands to spread bacteria.

This predipping and washing step not only serves the purpose of cleaning the teat, it also is the stimulus that causes the goat to release the hormone oxytocin, which in turn results in a letdown reflex that releases milk into the udder cistern. It takes anywhere from 20 to 60 seconds to get a letdown response, and it lasts 5 to 6 minutes. That's plenty of time to milk most goats.

There's more about the effects of hormones on milk production and release in the appendix, page 252.

Draw the Milk

Position the teat at the junction of your thumb and index finger, encircling it near the base of the udder. Though it's sometimes tempting on goats without a clearly defined teat, do not grasp the udder itself; that could cause udder tissue damage.

Squeeze your thumb and index finger together to trap milk in the teat. The teat will look something like a plump strawberry beneath your fingers. The closed teat must be held firmly; otherwise, when you squeeze the rest of the teat, the milk will be forced back up into the udder rather than out the orifice.

Next, gently but firmly bring pressure on the teat with your second (middle) finger, forcing the milk down even farther. The third finger does the same, then the little finger, and if all has gone well the milk has no place to go but out of the teat — not necessarily where you want it, on your first try, but at least out of the teat.

The first squirt or two from each teat (probably containing a drip or two of teat dip) should be directed into a strip cup, which is specially designed with a sieve or a black plate for a cover. That first stream is high in bacteria that have collected in the teat orifice and shouldn't go into your pail. In addition, use of a strip cup will enable you to see any abnormality in the milk, such as lumps, clots, or stringiness. This is an indication of mastitis, which demands your attention (see chapter 8). Never use the milk from any animal that's not in perfect health.

Once you have examined the milk on the lid of the strip cup, wipe the teats and your hands with a clean paper towel to remove any remaining teat dip. You are now ready to put the rest of the milk into your pail. Repeat the grasp and squeeze, alternating one hand on one teat and then the other hand on the other teat. Remember to open your fingers between squeezes so the teat can refill with milk from the udder. With a little practice you'll develop a rhythm.

Keep it up until you can't get any more milk, and then massage or gently "bump" the underside of the udder a few times, as kids do when sucking . . . although kids are not always gentle when they are eager for food. You'll be able to get more milk. This massaging is important, not only because the last milk is highest in butterfat but also because if you don't get as much milk as possible the goat will stop producing as much as she's capable of.

You never get all the milk out of the udder, but after several minutes you can tell you've reached the point of diminishing returns. At one time, the recommendation was to strip every drip of milk out of the teat by grasping it between forefinger and thumb and pulling down the length of the teat. The practice has fallen out of favor, both because there will always be at least a little more milk in there, and because it is too easy to damage the teat and udder with undue roughness.

When you're finished, dip the teat again, but don't touch it or wipe it off. The dip will

HOW TO MILK IN SEVEN STEPS

1. Close off the top of the teat with your thumb and forefinger so the milk flows out of the teat, not back into the udder.

2. Close your second finger, and the milk should start to squirt out. Discard the first stream; it will be high in bacteria.

3. Close the third finger. Use a steady pressure.

4. Close the little finger, and squeeze with the whole hand. Strive for a smooth, flowing motion. Don't pull on the teat; just squeeze gently.

5. Release the teat, and let it fill up with milk. Repeat the process with the other hand on the other teat.

6. When the milk flow has ceased, "bump" the udder, as kids often do while nursing, and you'll get a few more squirts. You never get all the milk, but it won't go away. It will be there the next time you milk.

7. Coat the teats with a teat dip to prevent bacteria from entering the orifice.

protect the end of the teat from bacteria for the 30-or-so minutes it takes for the orifice to close tightly. See the discussion under Mastitis on page 132.

I've never heard of teat dip harming nursing kids, especially in the tiny amounts that are left after milking. If you are worried, skip the dipping. The regular suckling action of the kid will do a good job of flushing bacteria from the teat end.

Machine Milking

Goats can be milked by machine, but we won't discuss this in a book for beginners. While some people do use milking machines even for just a few animals, others claim they can milk dozens by hand in less time than it takes to set up and clean a machine. And that's without considering maintenance, the cost, or the noise. For most of us, milking is the most peaceful and therapeutic activity of the day. Why spoil it with an air compressor?

Of course, there is always the problem of carpal tunnel syndrome, which was first known as "milkmaid's hand." Many sufferers have said they can successfully milk by hand simply by tucking their thumbs down into the

---

TIP TO PREVENT MASTITIS

Wash and dry your hands before and after milking each goat, or wear latex or nitrile gloves, which are easier to keep clean.

---

fingers and pressing the teat against the back side of the thumb instead of into the palm of the hand. It is a little awkward at first but puts less strain on the nerves in the wrist.

Problem Milkers

There are problem milkers. If you've learned to milk with decent animals, you can probably figure out how to cope with the other kind, but if you're unfortunate enough to have to learn on a troublemaker, some of the fun will go out of the experience.

First fresheners are most liable to be the culprits if they have not been handled often as babies, although older does sometimes develop ornery habits, especially when they know you're using them to practice on. First fresheners are also likely to have small teats,

---

HOW FAST TO MILK?

Some sources advise that the ideal milking speed for goats is 100 to 120 pulses per minute, whether milking by hand or by machine. This is faster than cows are milked, but it best matches the rate at which kids suckle and results in better milk letdown, these people say.

A 2008 document on best management practices for dairy goats based on many years of combined experience by the authors said the rate can range anywhere from 60 to 100 pulses per minute. Very few people can keep up a pace of 120 squeezes a minute for the length of time it takes to empty a goat. Go at a pace that is comfortable to you. The more important factor is that you not stop before the udder is empty. It takes much longer to establish a second letdown response, and you may not get all the milk out.

which makes milking difficult, especially if you have large hands. On some it's possible to milk by using the crotch of the thumb; others will require using the tips of the thumb and index finger in what amounts to stripping until their teats gain some size over time.

An occasional doe will tend to kick, and almost any doe might kick once in a great while. This generally indicates that something is wrong. She's bothered by lice or flies, or you pinched her, or your fingernails are too long. Placing the bucket as far forward as possible, away from her hind legs, will help in this situation, and you can lean into her leg with your forearm to control movement. Leaning into the goat with your shoulder and holding her against the side of the milking bench or wall will also serve to restrain ornery or nervous animals. It can also be useful in mild cases of "lying down on the job." Goats that have been nursing kids are especially prone to this sort of unhelpful behavior.

Milk Handling

Milk, especially raw milk, is highly perishable and extremely delicate. Following are some simple steps that will prevent spoilage.

- Cool milk immediately after milking.
- Don't add fresh warm milk to cold milk.
- Never expose milk to sunlight or fluorescent light for any length of time.
- Keep milk in the refrigerator, not on the table, until you need it for pouring.

TO COVER OR NOT TO COVER

Most "scientific" literature says it's absolutely necessary to keep the milk covered while it's cooling so it won't collect airborne bacteria. However, some folks swear that their milk tastes better when it's left uncovered while cooling or when the cover is left ajar. Cover it, and if your milk doesn't taste quite right, try it the other way.

A stainless steel pan is a good container for milking. You might even go one step further and invest in a hooded pan that helps keep dust, dirt, and hair out of the fresh milk.

Cooling milk immediately after milking means that it should not be left standing while you finish chores. Ideally, milk should be cooled down to 38°F (3°C) within an hour after leaving the goat. That's quite a rapid drop when you consider that it was over 100°F (38°C) when it left the udder. Home refrigerators aren't cold enough; generally, it takes an ice-water bath to do the job well (see Cooling and Storing, below).

Don't add fresh warm milk to cold milk. If you're accumulating milk for cheesemaking, develop a system for rotating it, perhaps from left to right, or one shelf to another, so you know which is freshest. Or you can just label the containers.

If you store milk in glass jars, be sure to never leave them in the sun or under fluorescent light, as this will change the flavor. Also, don't leave a container of milk sitting out after a meal in any event. Keep it cold.

Cleaning Your Equipment

All equipment that comes into contact with milk must be scrupulously clean. Cleaning milking utensils is quite different from ordinary household dishwashing. A dishcloth or sponge will not clean microscopic pores that hold bacteria that will spoil milk or give it a bad flavor: a stiff brush must be used. Household soaps and detergents contain perfumes that will leave a film on equipment and may cause off-flavors in the milk. Many household bleaches are not pure enough or strong enough for dairy work; you need special compounds. You shouldn't even use towels to dry equipment, because of the bacteria they contain.

There are four cleaning agents for dairy equipment. Two are for washing: alkaline detergents and acid detergents. Iodine and chlorine compounds are used for sanitizing, although sodium hypochlorite (Clorox brand bleach without added scents) is dairy approved.

The alkaline detergent is the basic cleaning agent. However, it leaves a cloudy film called milkstone, which harbors bacteria. To get rid of the milkstone, you must use an acid detergent, which doesn't have the cleaning power of the alkaline detergent. Most dairy farmers scrub their equipment with alkaline detergent for 6 days, and on day 7, when everybody else is resting, they scrub with acid detergent. If you have hard water, which hastens the development of milkstone, you can put the acid detergent in the rinse water every day.

COOLING AND STORING

If after reading "Raw Milk vs. Pasteurized" in chapter 2, you decide to pasteurize your milk, now is the time to do it . . . before it is cooled. There is more important information on page 227 about raw milk if you plan to make cheese.

Now back to cooling. For the best-tasting milk, cool it to 35°F (1.5°C) within 30 minutes after milking. Pour fresh milk through a paper milk filter into one or two sterilized glass canning jars. Twist on the lid, and set the jar(s) in an ice-water bath. About every 10 minutes, turn the jar gently upside down a few times to cool the milk evenly, and then set it back in the ice. Store in the refrigerator.

Measure all the washing and sanitizing materials carefully. If the solutions are too weak, they won't do the job they were intended for; if too strong, you're wasting money and you run the risk of contaminating your milk.

Here's the procedure for cleaning:

1. Rinse pails and other equipment thoroughly as soon as possible with cool water to remove the milk proteins. Warm or hot water will "cook" the casein in the milk, leading to a buildup that can be difficult to remove. Never let milk dry in the pail.

2. Mix alkaline as instructed, and scrub with a brush or nonabrasive woven plastic pad. Periodically, use an acid rinse according to instructions on the container.

3. Rinse in plenty of hot water, and let soak in a chlorine or iodine sanitizing compound, according to the time on the label. Clorox brand is approved for dairy use. Cheaper brands of chlorine are not.

4. Invert on a rack to air-dry. Do not dry with a towel. Grade A dairies, by the way, are required to store milk buckets and containers upside down, without covers, on wire racks, to allow exposure to air. Why not do the same in your home dairy?

5. For extra insurance against bacterial contamination, just before using the equipment again, soak it for 1 minute in a solution of 1 tablespoon (15 ml) of Clorox to 1 gallon (3.75 L) of hot water (100 parts per million available chlorine). Allow excess solution to drain away. Do not rinse or towel dry.

Commercial dairies must follow these procedures. The number of backyard goat raisers who go through all this is open to question. I've described the procedures not because you'll croak from drinking milk that wasn't produced under hospital conditions but so those who want to do a professional job will know they shouldn't be doing it with soapy dishwater and a dishcloth and towel. If you ever encounter "bad"-tasting milk, your milk handling and equipment cleaning procedures are the first things to examine.

PARTING SHOT

Even a home dairy with a few goats may produce more milk than you can use. If you want to sell excess milk to the public, be sure to understand and follow state law concerning fluid milk sales. Your county or parish Extension agents or state-level departments of health or agriculture are good resources. In some states, there are no regulations. In other states, the sale of fluid milk or cheese for human consumption is highly regulated or may even be outlawed unless it comes from a licensed dairy. There are very few places where any state law deals with the incidental sale of pet milk, however, except where you are required to pay tax on your income.

Barn Record

Date	Pokey a.m./p.m.	Mayo a.m./p.m.	Molly a.m./p.m.	Comments
1				
2				Clean Barn
3				Unusually warm
4				Melty - Muddy
5				New load of hay
6				
7				
8				
9				Snowstorm
10				Snow
11	Colostrum			Pokey 1B-1D easy
12	2 / 1.5			
13	2 / 2			
14	2 / 2			
15	1.5 / 2.2	Colostrum		Snow
16	2 / 2.2	2.5 / 2		Mayo 2 Bucks - easy
17	3 / 3.2	2.2 / 2.4	Colostrum	Molly 3 does - Pulled first
18	3.2 / 3	2.5 / 3.1	2.0 / 2.4	Watch Molly
19	3.8 / 3	3.4 / 3.1	2.0 / 2.0	Milked late a.m.
20	3.4 / 3.5	3.5 / 3.7	2.2 / 2.4	
21	3.7 / 3.8	3.7 / 3.6	2.3 / 2.5	
22	3.8 / 3.8	3.4 / 3.7	2.0 / 1.7	Molly depressed - separate pen p.m.
23	4.0 / 4.2	3.7 / 3.7	2.0 / 2.0	Molly back to normal
24	4.0 / 4.0	3.1 / 3.5	1.7 / 1.9	Vet - uterine infusion (Molly)
25	4.2 / 4.4	3.1 / 3.3	2.0 / 2.2	Pokey kids nibbling hay
26	4.3 / 4.4	3.8 / 3.6	3.0 / 3.1	Molly perking up
27	4.6 / 4.7	3.8 / 3.9	3.0 / 2.9	" Back on feed
28	4.4 / 4.8	4.0 / 4.2	3.0 / 3.2	Mayo kids doing great!
29	4.7 / 4.9	4.1 / 4.3	3.3 / 3.4	All back to normal
30	5.0 / 4.7	4.1 / 4.3	3.4 / 3.4	
31	5.0 / 5.1	4.2 / 4.4	3.4 / 3.5	Milk check - OK

Toss

14

KEEPING RECORDS

No livestock breeder or householder of any kind can afford to be without good records to use as a management tool. And any goat enterprise will have a much greater chance of future success if you know exactly what you've done in the past.

Record keeping is necessary for the commercial goat dairy, because only through accurate and complete records does the owner know if the operation is making a profit — and if not, why not.

It is a necessity for the show goat breeder because only accurate and complete records will help to upgrade goats to the hoped-for blue ribbon status.

Most homesteaders and other backyard goat raisers shun record keeping because they aren't involved with profit, or upgrading, or awards, and they think the work is a boring waste of time.

Big mistake. They're wrong on three counts.

It's true that the home dairy doesn't depend on goats for a living, as the commercial dairy does. But profits can still be realized in milk and dairy products that are better and cheaper than those purchased in the supermarkets. Even if the casual goat owner has no intention of ever entering a show ring or even getting near a goat show, it's still necessary to know certain facts about the herd's production and the results of management and breeding practices.

And record keeping can be fun! It becomes a challenge to have does that produce better than their mothers, and it's satisfying to look back on records that are several years old and see, in black and white, how you've progressed.

If you have registered goats, pedigrees and registration certificates will be an important part of your files. The person you buy registered goats from will help you get started and point you to the registry where your new goat is listed.

The Basic Barn Record

I talked about barn records in chapter 3 as a useful tool for selecting your goat. Here you will learn how helpful they are once you have your goat home. And who knows; your well-kept records may be the very thing to clinch a sale in the future.

The basic barn record is a chart showing how much milk each goat produces. While a monthly school calendar is handy (see Monthly Records, below), a plain sheet of paper works fine, with the goats' names written across the top and the days of the month down the left margin. You can write the morning's milk in

---

Keeping a daily record of milk production, health information, and changes in routine can go a long way to better understanding what is working for your farmstead and what changes might be helpful in the future.

one corner of each square or imaginary square, and the evening's milk below it, as 4.5/4, with the first number the morning's yield of 4.5 pounds and the second, the evening's of 4 pounds.

Milk is measured by weight rather than volume, for official records on both goats and cows. It's the best procedure for the home dairy, too. Freshly drawn, unstrained milk foams, and it's difficult to gauge actual production in quarts, pints, or even cups. Then, too, it's much simpler to deal in pounds and tenths of pounds rather than in quarts and fractions of quarts. For all practical purposes, a quart (1 L) of milk weighs 2 pounds (1 kg), and 8 pounds (3.5 kg) is a gallon (3.75 L).

It's a good idea to use this sheet to make notations of relevant data. For example, if you note "Susie in heat," you will be alerted to watch for her next cycle in 21 days. Notes on changes in feed; unusual conditions, such as a doe's not feeling well or acting quite right; or any other factor that might contribute to differences in milk production can be a big help in interpreting your records even years later. The weather can be important, too.

Information about individual goats — milk, breeding, and health records — are very handy to find all in one place, but a 12-month calendar could also be used, especially for giving a broad look at the whole herd.

Any medications used — vaccines or dewormers — should definitely be noted, including the type, amount, and date. And you might want to include hoof trimming and other routine chores on your calendar, too.

It's convenient to note breeding dates, the name of the buck, and freshening dates, with all pertinent information, right on this same sheet. When we bought feed instead of growing our own, that went on the sheet, too. At the end of the year, we had a complete record of the input, output, and interesting happenings in our barn, all on just 12 pages.

Recording Reality

One of the primary advantages of such a system is that it overcomes the natural forgetfulness of most human brains. Let's face it: few people, if any, are going to remember the statistics from 730 trips a year to the barn, which, if the herd consists of three goats with lactations of 305 days, means 1,830 separate entries each year for milk alone.

In addition to not being able to remember all those numbers, the brain can distort them. For example, you might be impressed by Susie's production of 1 gallon of milk in 1 day and consider her the best goat in your herd. But your records might indicate that a less spectacular producer that just chugged along less dramatically, but had a long and steady lactation,

MONTHLY RECORDS

It's important to keep a record for each animal, but many people like a barn calendar for that purpose. Particularly handy is the one made available to teachers, because it runs from August through July and makes it easier to keep fall breeding dates and spring freshening dates all in one place. Request one at the local school. They usually place orders in the spring or summer.

BASIC BARN RECORD

March	Calpurnia Milk Yield (a.m./p.m.)	Cleopatra Milk Yield (a.m./p.m.)	Nefertiti Milk Yield (a.m./p.m.)	Temperature/Comments
1	3/3.25			32° Cloudy and windy
2	3/3			32°
3	3/3			31°
4	3/3.25			33° First geese
5	3/3			33°
6	3.25/3.25			32° First robin
7	3/3.25			38° (Vet) Autumn disbudded
8	3/3			54° Thunderstorm
9	3/3			28° Snow!
10	3.25/3	—		20° Cloudy, 2 does!
11	3.25/3.25	—		25°, 100 lbs. feed
12	3/3	—		22°
13	3.25/3.25	2/1.5		21°
14	3.25/3.25	2/2		30° Gloomy
15	3.5/3.5	2/2		35°, ½" rain — ice off pond
16	3.5/3.5	1.5/2		11° Clear — ice on pond again
17	3.5/3.5	1.5/2		21° Overnight low 8°
18	3/3.25	2/2		32°, 1" snow, 6 bales hay
19	3.25/3.25	2/2		34°
20	3.25/3.25	2.25/2.25		35° Dreary, 13 chicks hatched
21	3.25/3.25	2.5/2.25		39° Foggy
22	3.25/3.25	2.5/2.5		45° Still gloomy
23	3.25/3.25	2.5/2.75		50° Light rain (at last!)
24	3.25/3.25	2.5/2.5	—	48°, 2 bucks
25	3.25/3.25	2.75/2.5	—	40°, 10 lbs. feed
26	3.25/3.25	2.5/2.75	—	40°
27	3/3.25	2.5/2.5	1/1.5	35°
28	3/3	2.75/3	2/2	32° Overcast, flurries
29	3/3.25	2.75/2.75	2/2	35° Cloudy
30	3/3.25	3/3	2.25/2.5	35°
31	3.5/3.25	3/3	2.25/2.5	35°

This is an example of a very simple barn record form. Real barn records are usually hastily scribbled, often stained, and wrinkled, and I've even had a few with goat teeth marks on them. The basic information is milk production, which is weighed and recorded at each milking. Other data such as feed purchases, medications administered, kidding dates, and even weather conditions can be included. To make these records even more valuable, you can transfer them later to a spreadsheet so you can determine the cost of your milk, create individual production records, and even chart lactation curves for each of your goats.

actually produced more than the flashing star. If you had to cull milkers or make decisions about whose daughters to keep, you might make the wrong choice without those records.

In many cases, breeders will note that a relatively few top does produce as much as a larger number of poorer producers. Since poor producers require just as much work as good ones and eat virtually as much, it follows that milk from the lower third or even half of your herd costs more than milk from the top half or two-thirds. This suggests that you could get more milk for the same amount of time, effort, and money by replacing the poor does with daughters from the best does. If you don't need that much milk, you might be able to eliminate one or more animals, reduce your feed bill, and still increase your milk production.

Records of breeding, expenses, income, and milk production are all basic, and it doesn't take much time or knowledge of accounting to keep them. Just the act of writing them down will tell you a few things about your operation, but it's also possible to squeeze a lot more helpful and interesting information out of those records. Today, with computers, it's easier than ever. Check the publications for programs written specifically for goats, or simply use a spreadsheet.

In any given category, your cost and income will vary according to your location, your type of operation, and your management methods. Even when these factors remain constant, your costs can vary from year to year. In a drought year, the price of hay can double. You might have mostly doe kids one year, with heavy registration expenses. The next year you might have mostly bucks that go to the butcher, and very low registration expenses. Medical expenses often come in spurts. And don't forget that if you produce more milk than your household can use and the surplus is used to feed calves or pigs, there should be a price differential.

Many people may be hesitant to pay a higher price for goat milk at the grocery store. Watching your costs is just as important at home by keeping good records.

THE BASIC BALANCE SHEET

Of course, you don't need a computer to tell you how much your goats are worth to you. You can keep monthly and annual tallies of a few simple income and expense items. Here's an example:

Income

Milk sales	$_____
Sales of stock	$_____
Family milk	$_____
Family meat	$_____
Stud fees	$_____
Boarding fees	$_____
Misc. (disbudding services, etc.)	$_____
TOTAL	$_____

Expenses

Purchase of stock	$_____
Feed (grain, minerals, salt)	$_____
Hay	$_____
Veterinarian and medicine	$_____
Repairs on equipment	$_____
Supplies	$_____
Advertising	$_____
Registry, transfers, etc.	$_____
Telephone, postage	$_____
Amortized costs	$_____
TOTAL	$_____

Profit (or loss) (income minus expense)

Note: This record can be much more detailed. You might want to divide "Sales of stock" into breeding stock, meat animals, and perhaps pets, to learn where your best market is. It could be very helpful to separate hay from grain. For example, you might look at your records, determine that you're spending too much on hay, and decide to build a new hay-saving manger. Later, you'll want to use those records to see how much money the new manger is saving you. And if it works really well, so the goats are no longer bedding themselves with wasted hay, you might want to add a category for straw!

Figuring Out Costs

It would be helpful to know in advance how much a gallon of milk from your home dairy will cost, perhaps as a way to justify the investment in a goat in the first place. Unfortunately, there is no reasonable answer to that. Even well-managed commercial dairies have costs that vary widely. Dairy plants that buy goat milk likewise vary in their payment schedules from place to place, and even month to month, as well as on protein and other factors. Recently, this has ranged from about $32 per hundred pounds (45 kg) to as high as $45. This is from $2.56 to $3.87 a gallon (3.75 L)! The cost to produce this milk varies just as widely in commercial herds and perhaps even more in home dairies.

That's precisely why you should keep records. Your own balance sheet is the only one that counts.

There's no point in looking at numbers and setting unrealistic goals for yourself — or being disappointed because you think you don't measure up. You should determine what your milk is costing you and balance that against what you're saving at the grocery store. But don't forget to add in the value of meat, fertilizer, the security and pleasure of providing your own dairy products, and the fun of having goats!

Putting a Value on Your Milk

Putting a dollar value on the milk from your home dairy is no simple, cut-and-dried calculation. Even if you go by the maxim that anything is worth only what someone else is willing to pay for it, goat milk presents special problems.

Most of us produce milk that has varying value. In winter, when production is likely to be low, the entire output might be used for drinking (the "fluid milk" market, the dairy industry calls it). If we'd be willing to pay the health food store price of fresh goat milk, this milk would be quite valuable. If you have a baby who's allergic to cow milk and you can't find goat milk — at any price — the cost of your home-produced milk is probably of little concern. If without goats we'd be drinking cow milk, it's somewhat less valuable. As the milk flow increases, we might begin to use some to make cheese or yogurt. This is "manufacturing" milk, and even cow farmers get less for it. When we begin to use an even greater surplus as feed for pigs, calves, or puppies, the value slides even further.

That's not all. Do you consider manure to be "waste," or is it black gold? Do culls go to the butcher, the rendering plant, or the person interested in a pet? Do you utilize the hides? In one way or another, all of these affect the cost of your milk.

In other words, you have a lot of leeway in putting a value on the milk you use yourself. Remember, you're not keeping these records or coming up with numbers for the bank or the IRS. These valuable tools are strictly for your own use, to improve your herd's performance.

Capital Costs and Operating Expenses

Figure your capital costs; that is, money you invested in things that aren't used up all at once. This includes milking equipment; feed pans and water buckets; fencing; tools such as the disbudding iron, clippers, and tattoo set; and the goat itself. Naturally, you don't want to charge all this against the milk produced in 1 year.

The milk pail might last 20 years: take one-twentieth of the price as this year's cost

THE MARKET FOR GOATS

Some sources suggest that many herds break even not because of the value of the milk but because of the value of the kids. A purebred and registered herd that can command top price for its animals will come out far ahead of a herd of grades whose kids are a drag on the market.

This is still true, but notice the wording: "can command top price." At a particular time in a particular place, even excellent purebred stock doesn't always bring high prices. This could be simply because there are plenty of good grades available, and that's what most people in the area want anyway. Still, many other factors could be involved. For top prices, not only do you need top animals, but you must earn a reputation in the show ring; you'll probably have to be on official milk test and have your animals classified, and you must advertise. All of these require time, money, energy, and often frustration. How can you know if it will pay off?

One way is to make projections, based on current prices and costs and market conditions in your particular area. This is just one more way of making records work for you. On the other hand, many goat owners simply aren't interested in increasing profits if it means going to shows, getting involved in registering animals and all that entails, or dealing with the public. Their records will instead emphasize the basics of simple costs in and simple expenses out.

(that's conservative: ours is going on 40; I have long forgotten what we paid for it). The goat might be good for another 5 years: take one-fifth of what you paid for her. Go down the list of capital goods, determine the capital costs for 1 year, and you'll have a more honest picture of your true costs.

Then add up your operating expenses: hay and grain, electricity used in the barn, veterinary fees, milk filters, and everything else that was purchased and used up.

Add up the operating expenses and 1-year cost of capital equipment and stock, and subtract that from the value of the goat's production. This will give you a pretty good idea of the goat's annual value to you. By adding up all these costs and dividing by the number of quarts or gallons of milk produced, you'll know the actual cost of your milk.

Even this isn't completely accurate, but it's adequate for most people and far better than a complete disregard for accounting. If you're inclined, you can figure in the cost (or value) of labor, the value of manure, the cost of borrowed money or the return on investment, taxes, and even more.

If more goat raisers kept such records and really knew what their goats were costing them, you can be sure there wouldn't be very many $10 or $20 goats for sale! More people would pay better attention to culling and proper management, too.

15

CHEVON

As with most backyard goat owners, it's easy for that one goat to become two, and then four, and so on until you hear your spouse comment to a friend, "I used to be able to tell you how many; now I just say, 'too many.'" One solution is to plan more meals around *chevon*, a gentile word meaning goat meat. If you're inclined to Tex-Mex cooking, you might call it *cabrito*, from the Spanish for "little goat," but except in the Southwest, *chevon* is the more common term.

The Market for Chevon

Goat meat by any name is very popular in many cultures. In the United States the popularity of goat meat used to be confined mainly to people of Spanish, Greek, and Jewish heritage, but Pacific Rim and African immigrants have greatly expanded the market. Kid is an important part of the meals for spring festivals of several religions, but chevon is a traditional — and delicious — everyday meat for many. Whole roasted kid as "cabrito" is making a splash in upscale markets.

Before 2002, the National Agricultural Statistics Service (NASS) of the U.S. Department of Agriculture reported meat goat numbers only from Texas, where goats could be raised cheaply on range and where large markets exist, and sporadically from New Mexico, Oklahoma, and Arizona. In 2006, those numbers became significant enough across the country that NASS began doing a twice-yearly survey. In 2016, there were more than 2.1 million meat goats in the United States and another 375,000 milk goats.

Even as far back as 1997, an estimated 30,000 goats a month were shipped into New York City. More than 60 percent of the goat meat sold there was purchased by Muslims. But almost every region has at least a small niche market for goat meat today. This is one reason for the soaring popularity of meat goats such as the Boer. Boers and Boer crosses grow much faster and are "meatier" than dairy breeds.

Domestic production of goat meat has increased steadily since those early 2002 figures, when 595,000 goats were slaughtered at federally inspected meat plants. In 2007, that number was 638,000 goats, and in 2009, it was close to 650,000. Despite the rising numbers in domestic production, there apparently was not enough to meet the demand, because the United States was importing frozen goat meat from Australia and New Zealand. The USDA reported in 2014 that the United States imported 43,188 million pounds (19,590 million kg) of goat meat.

While buck kids are as much fun to watch as the doelings, fewer than one in a hundred will be worth keeping for breeding purposes. Waste not, want not, as the saying goes. Plan ahead to use those bucks in your farmstead food plan as chevon.

Goat meat can be prepared dozens of ways for healthful, colorful, and flavorful meals.

What Does Chevon Have to Do with Dairy Goats?

Meat is an important by-product of dairying. Over the years any farm will average 50 percent buck kids. Not one in a hundred can be kept, profitably, as a herd sire. While there is a limited demand for wethers (castrated males) as pets in some areas, it is more merciful in most cases to butcher them for meat.

In addition to unwanted males, any dairy operation will have cull or aged does that simply are not paying their way. Resist any temptation to sell them as milkers to someone else. You might make a few dollars on the deal, but the cost to the goat world — and to your reputation — will be far higher.

Culling is a fact of life when dealing with livestock, but that doesn't make it any easier — especially for city people with no livestock experience and even more so when dealing with endearing animals like goats! Butchering is never a pleasant task, and it's normal to have qualms about eating an animal you raised yourself. Native American and other cultures move beyond the philosophical dilemma by honoring the life of the animal before taking it for human nourishment. Once you overcome any initial reticence, by whatever means works best for you, you'll most likely agree that chevon is a delicious bonus of your home dairy.

Slaughtering and Butchering

Goats are commonly slaughtered at one of four stages:

- On farms where milk is valuable or where the labor required to raise kids is deemed out of proportion to the value of the meat, kids may be butchered at birth and dressed like rabbits.
- Milk-fed kids weighing between 20 and 30 pounds (9 and 14 kg) are the most

popular for the Mexican dish cabrito and for the Easter-Passover market.

- Castrated buck kids may be kept and butchered at 6 to 8 months for a goat-meat market or your home table.
- Cull does can be processed into ground meat, jerky, salami, or anything that makes use of meat that isn't especially tender.
- A fifth category should be mentioned as well. For some ethnic markets, very smelly, very rank bucks are preferred.

Dispatching the Animal

In many places, a local meat processor or locker plant will slaughter, cut, wrap, and freeze a goat for you. The cost of full-service processing depends entirely on where you live and whether the abattoir is interested in dealing with small ruminants. In many cases, the cost of custom processing negates any savings you might have accumulated by growing your own animal, but it solves the problem of dealing with the unpleasant job yourself. You take the animal to town on the hoof and bring it back in neat little packages.

However, if you are truly on a path of self-sufficiency, you'll need to know how to do the whole job on your own. At the processing plant, the animal will probably be humanely killed using a penetrating captive bolt or stunner that enters the brain with a fast and

---

GOAT MEAT BY ANY OTHER NAME

In Sir Walter Scott's *Ivanhoe*, Wamba the jester gives Gurth the swineherd an English lesson. While domestic animals are alive and must be tended, he explains, we use their simple Saxon names: cow, calf, sheep, pig, and so on. But when they are dressed and served to the Norman conquerors, they are called by their Norman (French) names: beef, veal, mutton, and pork.

In twentieth-century America, the equivalent of the Norman conqueror was the consumer. People weren't very interested in eating goat, so goat farmers once again drew on the French to create a word they hoped would be more appealing: *chevon*, from the French words for goat (*chèvre*), and sheep (*mouton*).

---

decisive blow. At home, the same end can be achieved by dispatching the animal with a gun. It is also fast and humane if the projectile is well placed. The front of the head is occasionally too hard for a clean shot. You may wish to aim from the back of the head, both to get better penetration and so as not to frighten the animal. Hold the gun roughly 2 inches away from the animal so the bullet has time to pick up spin and speed leaving the barrel.

Dispatching meat kids can be done the same way, but some people view it as a waste of ammunition. Even Temple Grandin, nationally known expert on humane handling of livestock, has agreed that a decisive blow to the back of the head with a sturdy tool is effective and similar in effect to the commercial captive bolt.

SLAUGHTERING TIP

Withhold feed, but not water, for 24 hours before slaughtering. Butchering will be easier and cleaner if there is no feed in the paunch.

Hanging and Bleeding

Although some people claim it's merely a cultural practice, most will say that thorough bleeding is important for both appearance and taste of the finished meat. As soon as the animal is stunned, cut through the animal's gullet and be sure to sever the carotid arteries on both sides of the neck. Do not damage the heart in the killing process, so it can continue to pump blood. Hang the animal head down to allow complete drainage. If you do much butchering, a gambrel hook will prove useful, but a carcass can be hung by passing a metal or wooden rod through the tendons of the rear legs or even by tying it, with legs spread, to a rafter or a tree branch with a rope, as many deer hunters do.

If you are, or know, a wild game hunter, you might have access to a gambrel hook like this one. But if you're only going to butcher one or two goats each year, you can make do with a piece of wood such as a tree branch to hold the hind legs apart. Or simply suspend the carcass with ropes, again spreading the hind legs.

Skinning

We've been told that the Greeks, who have a long tradition in goat butchering, cut a small incision between the hind legs and blow up the hide like a balloon. This helps separate the hide from the meat and makes skinning easier and cleaner. No doubt most people would rather use a tire pump than their mouths, but one goat raiser reported inserting the nozzle of a garden hose into the incision and filling the space with cold water. In addition to separating the hide, this helped to cool the meat.

To skin the animal with or without this step, carefully cut a slit through just the skin layer from between the hind legs to the throat. Don't cut too deeply, as you don't want to cut into the thin stomach wall or meat. Once the cut is started, you can usually work your fingers beneath the skin to hold it away from the body. Cut off all four feet at the joint above the pasterns.

From the two ends of the central cut, continue out along the insides of all four legs. Be aware that the skin is tighter on the legs, and again, try not to cut into the meat.

The pelt will be attached at the tail. To remove that, cut around the anus and loosen it until you can pull out a short length of colon. Tie off the colon with a piece of strong string to avoid possible contamination, and let it fall back into the body cavity. Then cut off the skin at the base of the tail. If you're very frugal, skin out the tail and use it for stew meat. Carefully, peel the hide downward off the legs and body. At this point, you should not have cut into the actual body cavity anyplace but the anus.

If you're saving the hide, cut it off as close to the ears as possible. Skins from newborn goats are more like fur than hide, and many useful items can be made from any tanned goat skin. Remove the head from the carcass at the base of the skull.

Butchering

After the hide is removed, make a shallow cut down the belly, from between the hind legs to the brisket. If the animal was without food for 24 hours before slaughter (again, don't withhold water), the paunch will be empty and there will be less chance of cutting into it, but be careful anyway. Let the paunch and intestines roll out and hang.

Work the loosened colon end down past the kidneys, and carefully remove the bladder. Some people cut through the pubic arch, or front of the pelvis, at this time. Pull out the liver, and remove the yellowish gallbladder from its surface by cutting off a piece of the liver with it. If the gallbladder breaks and spills bile on the liver, wash the meat in cold water immediately to avoid a bitter taste.

Reach down into the chest cavity and find the base of the gullet. Once it is cut free, the offal will fall loose.

Saw the brisket to open the rib cage — an ordinary carpenter's saw will work if you don't have a meat saw — and remove the heart and lungs. Cut through the front of the pelvis if you haven't already. Clean out any remaining pieces of tissue, wash the entire carcass with cold water, and wipe it dry.

If you like the organs, the skull can be split to get the brains and the tongue can be removed through the bottom jaw. Wash the liver, heart, and tongue in cold water, and drain them.

Cutting, Dividing, and Packaging the Meat

Newborn goats weighing about 8 pounds (3.5 kg) can be cut up like rabbits. Cut through the back just in front of the hind legs and again just behind the front legs. Each of these can be cut along the spine, giving you six pieces of meat.

Larger animals, cut like lamb, will yield roasts, chops, ribs, and trimmings that can be ground and used in patties or mixed with pork for sausage. Some of the larger pieces, such as the legs, can be corned or cured like hams.

If you have little or no butchering or meat-cutting experience, the thought of converting a carcass into neat packages like you see at the meat counter might seem like a formidable task. But don't let it throw you. Simply cut off pieces that "look about right to be a roast," for instance, and they'll be just as edible and probably even better tasting than those from the supermarket. You really can't do anything "wrong" at this point.

IDEAS FOR COOKING CHEVON

If you enjoy lamb, use your favorite lamb recipes with chevon. Since we raise both sheep and goats, we often have both in the freezer, and it's hard to tell the difference.

There are many ethnic dishes from such areas as Greece and Turkey that call for chevon. Oregano is a good spice to use with chevon, and the meat is excellent in curries. You'll find a few suggestions for getting started with chevon cookery in chapter 17.

16

DAIRY PRODUCTS

Great-tasting fresh milk is the primary reason for owning a dairy goat. But since goats are like potato chips (you can't have just one), and because of the fluctuations in milk production, sooner or later you're going to have more milk than your family can possibly drink. Then it's time to think about other dairy products.

When I started with goats, yogurt was unheard of in our neck of the woods. Today it's practically a staple, and many people make their own from store-bought cow milk. In the past decade, goat cheeses have become immensely popular as well. So you might have thought about those even before getting your first goat. Even then, you might not yet be aware of all the possibilities.

One of the joys of having goats is the dairy products you can make in your own kitchen. At certain times of the year, you will have a surplus of milk that can be turned into a variety of products that will make you more independent of the supermarket and will make your goats more valuable to you. It doesn't make sense to produce wonderful milk and then throw half of it out because you can't drink it all. And perhaps most important of all, making cheese and other dairy products is satisfying and a lot of fun!

Preserving Milk for Future Needs

When your goats are dry, it certainly isn't fun to have to buy hay and grain for them and milk for yourself. If you know you're going to run out of milk at some point, you'll probably want to consider preserving some fluid milk when there is excess.

Frozen Milk

Freezing milk is simple. This can also be a good idea if you don't have enough to bother making cheese and already have enough yogurt.

Freeze it in plastic jugs, leaving an air space for expansion. Thawed frozen milk is somewhat watery. While it's fine for cooking, you might need to give it a quick zip in a blender to remix any of the solids that may have separated out to make it more suitable for drinking.

Canned Milk

Although it's controversial, milk can also be pressure canned. Home economists say that canning milk at home is dangerous. Customer service representatives at several companies that manufacture pressure-canning equipment have told me they had no information on canning milk and that you can't can milk

---

Bright white milk is only one of the great dairy products that can come from your kitchen when you have goats. With simple recipes and ingredients, you can turn your milk into yogurt, butter, and a wide variety of tasty, traditional cheeses.

because "it will curdle." Some people who have been canning milk for years say this is all nonsense. You'll have to make up your own mind, but should you decide to try it, here's how:

1. Follow your regular pressure-canner instructions regarding the amount of water to use, allowing steam to escape before closing the vent, and so on.

2. Fill sterilized canning jars to within 1 inch of the top with fresh, warm (120°F [50°C]) milk.

3. Add sterilized lids and rings.

4. Process at 10 pounds' (0.69 bars) pressure — 25 minutes for quarts and 20 minutes for pints.

5. Let the jars cool in the canner, undisturbed.

6. Remove cans, and store in a dark place.

Canned milk has a slight caramel-like flavor, so it's best used for cooking and baking. Also, the butterfat comes to the top and the calcium settles to the bottom in canned milk. Just shake the jar before you open it.

It is possible to can milk in a boiling-water bath, but it is definitely not recommended, and the safety of the final product is in doubt, even if you plan to only feed it back to your goat kids. In fact, it is nearly impossible to find anyone willing to put out a recipe for this method. Commercial dairies that package ultra-pasteurized milk suitable for storage without refrigeration heat the milk to 275°F (135°C) for a few seconds, but you won't be able to get your water hotter than the boiling point where it turns to steam and fogs the windows. If you have enough time and water to

boil for 3 hours with the hope you have killed all the bacteria in your milk, be my guest, but it is far safer and much more economical to freeze it or pressure can.

Evaporated and Condensed Milk

If milk is heated to about 190°F (90°C) and then simmered very slowly, the water will evaporate. When it's reduced by half, it's evaporated milk. Simple. The kicker is that this can take up to 2 days. If you want to try it, probably on a woodstove and when you can use the humidity, use a large double boiler or one of those insulating pads that provide an air space between the heat and the pot to prevent scorching. For larger quantities, improvise a double boiler by putting about 2 cups (0.5 L) of milk in each quart jar. Place the jars in a canning kettle containing about 2 inches (5 cm) of warm water, and add enough water to reach the level of the milk. Keep adding boiling water as the level drops.

Condensed milk is easy to make with cow's milk, because you can increase the solids by adding powdered milk. Goat milk is more problematic, because it's hard to come by powdered goat milk. To make sweetened condensed goat milk at home, add 2 cups (0.5 L) sugar to 2 cups milk and reduce it by cooking. One cookbook author advises only 1 part sugar to 2 parts milk and still calls it "excessively sweet." She also says that if the milk is simmered, stirring often, it will be ready in about 2 hours.

Dried Milk

Most people say you can't dry milk at home. After investigating the commercial process, I agree. Even Mary Bell, who wrote *Mary Bell's Complete Dehydrator Cookbook* and dries such far-out things as pickles and watermelon, doesn't dry milk.

However, Mary Jane Toth's *Caprine Cooking* passes on a method furnished by one Jill G. Simkins, who has done it successfully. Here's the method: She simmers 1 to 2 gallons (3.75 to 7.5 L) of milk in a double boiler until it's evaporated to the consistency of cream. She pours that into a large pan and dries it in the oven, with the door ajar, at about 150°F (65°C) until the rest of the water is evaporated and the remaining product turns crusty. The dried product is then ground in a blender. To use it, she soaks 1 part milk in 4 parts water.

Cajeta

As long as we're talking about cooking down goat milk, it's a good time to introduce *cajeta* (pronounced *cah-HAY-ta*), a popular Mexican dessert topping. It also makes a great fruit dip similar to caramel but with more attitude.

Stir together 2 quarts (1.9 L) of goat milk, 2 cups (0.5 L) of sugar, 1 cinnamon stick, and 1 tablespoon of vanilla extract (you can use a vanilla bean split in half if you prefer). In a very deep pan, bring the ingredients to a boil. Remove the mixture from the heat, and whisk in 1 teaspoon of baking soda dissolved in a little water. It's going to get very frothy, but keep stirring. Bring the mixture back to a simmer (not a rolling boil!), and stir continuously.

This is the time to have a good book on hand, because you will be stirring for at least an hour, until the milk turns a golden brown. Remove the cinnamon stick and vanilla pod, if you put one in. Keep stirring to keep the milk from sticking to the bottom of the pan as it thickens and gets darker over the next 15 minutes. It is ready to pour into clean, warm glass jars when it coats the back of the spoon.

It will get even thicker as it cools. If it gets too thick, you can stir in a little water. Store in the refrigerator.

Cheesemaking

Sooner or later you'll want to try making cheese. In its most basic forms, cheesemaking doesn't require much equipment, takes little actual working time, and can become quite a hobby or even a business. There is as much art and science to making cheese, however, as there is to making wine. Don't expect to come up with any "vintage" cheeses without experience or some luck, but you will certainly come up with some fun and tasty favorites.

But remember that the very first cheese was probably a "mistake." The theory is that some ancient unsung shepherd carried milk for his lunch in a pouch made from a kid's stomach, which is where cheesemaking rennet comes from. The milk formed a curd, which the shepherd was either curious, hungry, or dumb enough to taste. It isn't provable, but it makes a good story.

Whatever the origin, what has evolved are many hundreds of supreme cheeses, virtually all developed and perpetuated by rural homesteads, and probably as many by accident as by design. Modern technology only tries to imitate those homestead cheeses.

Until recently, however, we homesteaders usually confined our cheeses to a very few, very simple types, mostly soft, fresh cheeses. Aging or curing is an additional step that requires more interest and dedication and a little more equipment. Today, thanks to a number of people who have taken home cheesemaking to new heights, almost anyone can make dozens of specialty gourmet cheeses at home. Many mail-order companies now supply everything you could possibly need in the cheese department and lots more. Yes, you can make not only cheddar, mozzarella, feta, and brick but also several forms of chèvre, Gouda, queso blanco, and hundreds of others, including many that gourmets rave about.

The Basics of Cheese

In its simplest forms, cheese is nothing but curdled milk with the whey drained off. But starting with this, well over 500 named cheeses can be made.

The milk can be curdled with animal rennet (an enzyme), vegetable rennet, acids such

Several varieties of soft French-style cheeses and cream cheese are made by hanging the soft cheese curd in butter muslin until the whey drains off. Note that the fabric is a tighter weave than what is commonly referred to as cheesecloth.

as vinegar or lemon juice, or with a combination, each producing different results. Curdling requires heating, but a few degrees' difference in temperature can affect the end result, as can the length of the heating period. Starter cultures can be added, the specific bacteria in each culture contributing still more differences. Some cheeses are pressed; others are not. Those that are can be held for variable times, under variable pressures with differing amounts and methods of adding salt. The temperature and humidity in the kitchen, the size and shape of the cheese, if and how it is aged — all this and more determines what kind of cheese that basic curdled milk will become.

Oh, and don't forget the milk! The composition of the milk can be a factor in how your cheese turns out, but the most important thing to remember is that you can't turn bad milk into good cheese. Always use fresh, good-tasting milk.

As you can see, cheesemaking can be much more complicated than goat raising. If you really get into it, you'll want to study books and articles devoted to it, search the Web, perhaps take classes, and network with other cheesemakers. You will find several, and sometimes many, recipes for cheeses with the same name but slightly different procedures, and it's likely that each will taste different. Experience will tell you which ones are best for you.

There are two important things to know about goat milk when it comes to making cheese and other fermented products. The first is that the small fat and protein particles make a curd that is very soft and intolerant of rough handling. If you stir or pour hard, the curd will break and give you very little cheese for your efforts. The second thing is that goat milk that is heated above roughly 175°F

(80°C) will denature — that is, protein chains will unravel — and it will not form a good curd when combined with rennet. Cheese coagulated with acid, though, will be fine.

There are several good books that have been written for homestead cheesemakers,

TO PASTEURIZE OR NOT?

You should be aware that the use of pasteurized milk in cheesemaking is another hotly debated topic, although with nuances slightly different from the raw-milk-for-drinking controversy. Many modern recipes call for pasteurized milk, which means that the naturally occurring bacteria, good as well as bad, have been killed. Then the cheesemaker adds specially chosen bacteria to produce a certain kind of cheese. This provides more control over the process and end product, but raw milk can produce more exquisite cheeses that even peasant (as well as snobbish) gourmets rave about.

It's interesting to note that after 2 months of aging, cheese develops antibiotic properties that kill almost all disease germs that might have been in the milk. I recall reading a news item several years ago about a cheese, probably goat, found at an archaeological dig, that was said to be several thousand years old and was still edible. Don't ask me how they knew it was edible!

How you want to proceed is up to you, but a good rule of thumb is that fresh cheese that requires long setting times without refrigeration is safest if made with pasteurized milk.

including ones with recipes specifically for goat milk. But just to get you started, here are a few simple recipes to try.

Basic Cheese

You can make several types of basic cheese with equipment you probably already have in your kitchen, plus a dairy thermometer. A large stockpot, stirring spoon, colander, and cheesecloth, along with lemon juice or vinegar and, of course, some milk, will get you started. Note that the kind of cheesecloth I'm talking about is not the gauzy stuff found in the hardware store. The proper material is more accurately called butter muslin, has much smaller openings in the weave, and can be found in a fabric store. Wash it in mild detergent and a little chlorine bleach to remove any debris before using it for cheese, and *always* sanitize your cloth just as you would sanitize other surfaces that come into contact with food. A proper sanitizing solution for all your cheesemaking equipment is 1 tablespoon Clorox to 1 gallon of hot water.

For a metric conversion chart, see page 278.

Vinegar Cheese

Milk that is several days old often works best with this recipe.

    2 quarts milk
    ¼ cup vinegar, lemon juice, or lime juice

1. Heat the milk to 185°F (85°C), or close to boiling. Remove from heat.
2. Stir in the vinegar or juice, and set the milk aside, covered, for 15 minutes. A layer of curd should form and separate from the greenish whey.

3. Gently ladle the curd into a colander lined with butter muslin, or gently pour the curds and whey into the colander (this is easier, but skimming results in a better flavor and texture).
4. Tie the corners of the cloth together, and hang the cheese where it can drain for several hours, or until it stops dripping.
5. Slice or cube, and eat it as is, although it's rather flavorless. Or make into queso blanco (see recipe on page 230.) For more zip, mix in fresh herbs or a favorite spice.

Ricotta

Make ricotta by following the recipe for vinegar cheese above, but use fresh whey instead of milk and bring the liquid up to 200°F (93°C). The whey, of course, is left over from making other cheese: you get two for the price of one! Your yield will be much lower, because you have already removed most of the solids in the first processing.

Cottage Cheese

To make cottage cheese, you replace the acid with rennet, change the heating procedure a little, and treat the curds a bit differently. The resulting cheese is delicious as is, or add salt, cream, chives, or whatever you like. At some point, you may want to make your cottage cheese with a bacterial starter, which gives it a different flavor, but this is a good beginner recipe. It should keep about 1 week in the refrigerator, but once set on the table, ours never lasts nearly that long.

1 gallon goat milk

¼ tablet rennet (not to be confused with junket)

½ cup cold water

1. Warm the milk to 86°F (30°C) in a double boiler or over some other indirect source of heat.

2. Crush the piece of rennet tablet with a spoon, and dissolve it in the cold water (or follow the directions that come with the rennet you use, liquid or tablet).

3. Add the rennet solution to the milk, stir slightly, and let it stand in a warm place until a curd forms, usually about 1 hour.

4. Test to see that the curd "breaks" by inserting the dairy thermometer or other object 1 inch or so into the curd, lifting it out, and making sure the curd breaks with clean edges.

5. When the curd breaks, cut it into small cubes with a long, thin-bladed knife (a); a serrated bread knife works very well. Make a graphlike pattern by cutting the curd vertically (all the way through to the bottom of the pot) into ½- to 1-inch squares (b). Then slant the knife about 45 degrees, and make slices at right angles to the first ones (c), reaching to the bottom and sides of your pot (see illustrations at right).

6. Let the mass sit undisturbed for about 5 minutes, then begin stirring very gently, and cut any large pieces remaining. If you like small-curd cottage cheese, use a wire whisk to gently cut your curd into small pea-size pieces.

7. Warm the curds and whey to 110°F (43°C), over about 30 minutes. Within limits, the longer it takes to get to this temperature, the firmer the curd will be.

8. Pour the curds and whey into a colander lined with butter muslin, and drain for a few minutes.

9. Run cold water over the curds and whey to rinse off the whey. This gives it a milder flavor and increases its shelf life.

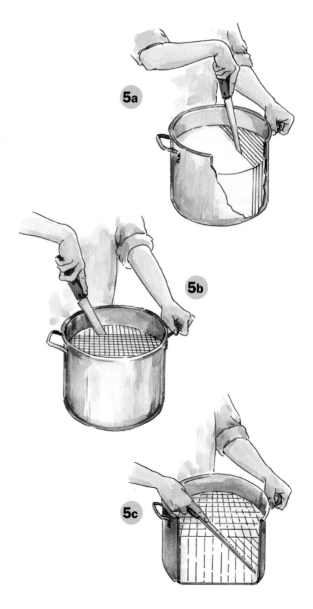

Goat Milk Cheddar

This cheese needs a mesophilic (medium temperature) bacterial starter, which you can find at a cheesemaker's supply shop or on the Internet. While you are there, look into the options for liquid microbial rennet, which is much easier to use than a rennet tablet. Working in a double boiler–type combination of pans allows better control of temperatures. Use a cheese press, if you have one, or use whatever clean weight you can find.

2 gallons fresh goat milk

1 packet direct-set mesophilic starter

¼ rennet tablet diluted in ¼ cup (60 ml) cool water (or 1 teaspoon (4.9 ml) liquid rennet in water)

2 heaping tablespoons cheese salt

1. Heat the milk to 85°F (30°C), and thoroughly stir in the starter. Allow to sit covered for 30 minutes.

2. Add the diluted rennet, and stir thoroughly. Hold at 85°F for 1 hour until curd forms and begins to draw away from the side of the pan.

3. Cut the curds in ½-inch cubes, and let them rest for 10 minutes.

4. Begin to very slowly raise the heat to 98°F (37°C) over 30 minutes. Stir gently to keep curds from matting.

5. "Cook" at 98°F for 45 minutes, stirring gently.

6. Drain off the whey, and stir the salt into the warm curd.

7. Drop curds into a large square of butter muslin. Twist the ends of the cloth to make a tight bag. Place it on a drain board with 20 pounds of pressure on top of it for 15 minutes.

8. Open the bag, turn the cheese over, twist the bag, and press with 30 pounds of pressure for 1 hour.

9. Turn the cheese again, and press for 2 hours under increased pressure.

10. Remove the cloth, and rub the cheese with additional salt. Turn and rub with more salt once a day for 2 days.

11. The cheese can be eaten young, but it is best if aged 4 to 12 weeks at 50 to 55°F (10 to 13°C).

Queso Blanco

This Mexican "white cheese" is a slight variation on vinegar cheese but has the unusual property of not melting. You can use it in stir-fry dishes in place of tofu, or fry 1-inch cubes in hot oil and serve with a yogurt dip for a special snack. Cubes can also be marinated in olive oil, wine, soy sauce, or any concoction that pleases your palate. You can also mix in pimientos, green chiles, jalapeños, olives, and salt to taste just before hanging the curd to drain.

1. Heat 1 gallon of fresh goat milk to 185 to 190°F (85 to 88°C), stirring constantly. Do not boil.

2. Slowly stir in ¼ cup vinegar until the curd starts to form and separates from the whey.

3. As with the vinegar cheese, ladle the curd into a colander lined with butter muslin, and hang until the whey stops dripping.

4. For a product firm enough to slice, twist the butter muslin tightly, and set the cheese on a cookie sheet in a shallow pan. Put a weighted plate on top of the bag and press the cheese for 8 to 10 hours.

Gouda

An easy hard cheese to start with, suggests Ricki Carroll of New England Cheesemaking Supply, is Gouda, named for the region in Holland where it originated. Gouda can be enjoyed after only 2 months of aging, but it develops a superior flavor after 6 months or longer. A cheese mold — in this case, the container that gives a cheese its shape — is needed for this cheese. You can also use it for the cheddar already described.

2 gallons goat milk

6 ounces buttermilk culture, made 1 day in advance by sterilizing 1 quart goat milk for 30 minutes in a water-bath canner, cooling to room temperature, adding a packet of Hansen's buttermilk culture, and setting for 12 to 24 hours

Few drops liquid cheese coloring

½ teaspoon liquid microbial rennet

½ cup cool water

1. Heat the milk to 86°F (30°C).
2. Add the buttermilk culture. Keep the milk at 86°F for 30 minutes.
3. Add the coloring, unless you don't mind a Gouda that looks somewhat like lard.
4. Add the rennet to the cool water, and gently stir this solution into the 86°F milk. Continue to maintain this temperature for 30 to 45 minutes. (Energy-saving tip: Put the cheese kettle in a larger pot or sink full of water that's about 86°F.)
5. Check that the curd is firm and has a "clean break": insert the dairy thermometer into the curd at an angle and see if the curd breaks cleanly. If it does, it's ready to be cut.
6. Cut the curd into fairly large 1¼-inch pieces, as described on page 229 in the cottage cheese recipe.
7. Let the curd set for a few minutes, until it sinks into the whey.
8. Ladle or carefully pour whey out of the kettle until the curds are visible.
9. This is a step peculiar to Gouda: add a quantity of water heated to 175°F (80°C), until the curd-whey-water temperature reaches 92°F (33°C). The amount of water isn't important; it's the temperature of the curd that counts. Let it sit for 15 minutes, stirring often. The curds will become smaller and harder.
10. Let the curds settle to the bottom, drain off the whey until the curds are visible, and add more water at 175°F. This time bring the temperature of the curds and whey up to 96°F (35°C), and maintain that temperature for 30 minutes.
11. Drain the curds, and quickly place them in a butter muslin–lined mold. Press with 30 pounds of pressure for 15 minutes.
12. Remove the cheese from the mold, turn it over, and put it back into the lined mold. Press for 6 hours with 50 pounds of pressure.
13. Remove the cheese from the press, remove the cloth, and let the cheese set in the mold overnight without pressure.
14. The next day, make a brine by adding salt to a gallon of water until no more salt will dissolve. Soak the cheese in the brine for 2 to 3 hours.
15. Air-dry the cheese until the rind is dry to the touch. This can take 1 to 3 days.
16. Wax the cheese. Age for up to 12 months at 40 to 50°F (4.5 to 10°C). You can serve it sooner, but age at least 60 days if you used raw milk. Turn once a day for the first several weeks and once a week after that.

HOW TO WAX CHEESE

Waxing "seals" cheeses that are to be cured, keeping out air. Although ordinary paraffin can be used, you'll get better results with a cheese wax, available from cheesemakers' supply companies.

1. Melt the wax in a double boiler to reduce the danger of fire. The wax should be fairly hot, but not so hot that it would burn a finger quickly dipped into it a little way.

2. Dip a cool, very dry cheese into the wax, and immediately remove it. You don't want the cheese to warm up.

3. When this thin layer cools and solidifies, dip another section of the cheese in the same way.

4. Continue this process until you have an even coating of wax about $\frac{1}{16}$ inch thick.

CHEESE WAXING TIPS

- If the cheese is too moist, the wax won't stick.

- If the wax forms a thick layer, it's probably too cool.

- If you don't get a definite layer of wax, then the wax is too hot.

Feta

Feta is a salty Greek cheese that is served crumbled over salad greens, as a pizza topping, or in casseroles and pasta dishes.

1. Follow the recipe for cheddar cheese through step 3.
2. Gently stir the cheese for 20 minutes at 86°F (30°C).
3. Pour the curds into a colander lined with butter muslin. Tie the corners, and hang the cheese to drain for 4 hours.
4. Untie the bag, and cut the mass into 1-inch cubes. Sprinkle the cubes with salt to taste. Age for 4 to 5 days in the refrigerator in a covered bowl. For a stronger flavor, store the cubes in a brine solution of ⅓ cup salt and ½ gallon water for 30 days in the refrigerator.

The possibilities for cheesemaking are endless. True, there are certain cheeses that can't be duplicated exactly at home, including Limburger, Camembert, and others requiring special cultures that are closely guarded secrets. Some cheeses require special climates or particular caves for aging. But there are more than enough recipes and cultures to keep any cheese-loving goat owner happy — and very, very busy.

Using Whey

You don't have to discard the whey you ladle, pour, or drain off. You can make ricotta cheese with it (see note on page 228) or the dark brown cheese known as gjetost. Whey also makes good feed for chickens and pigs. Don't feed it to goats, either kids or adults; it will cause scours.

You can make an interesting "lemonade" with whey, as described in a nineteenth-century cookbook. Here's how: Strain 1 quart of whey. Add 6 tablespoons of sugar or honey, the juice of 2 lemons, and a dash of nutmeg or cinnamon. Serve chilled.

Yogurt

Yogurt has been eaten and cooked with for centuries throughout the Balkans, the Middle East, and India, where goats are the standard dairy animals. The older spelling, yoghurt, is directly from the Turkish. Homemade yogurt is so superior to the supermarket variety that there's no comparison. Incubation methods range from warm-water baths to commercial yogurt makers, to heating pads, to solar energy. There are so many ways of using yogurt. Try it, for instance, as a substitute for sour cream or light cream. In Middle Eastern cooking, it's used as the liquid in many kinds of stews and soups. Goat-milk yogurt doesn't curdle when cooked, as cow-milk yogurt does. It also tends to be a thinner consistency and is often referred to as "drinkable."

Making Yogurt

To thicken the milk you will need a yogurt starter or culture. These starters are considered "good bacteria" and are packaged individually or in combination, depending on the supplier. The typical starters are *Lactobacillus*, *Streptococcus thermophilus*, and *Bifidobacterium*. You can buy dried cultures from a cheesemaking supply business, or you can go to the grocery store and buy any unflavored yogurt with "live active cultures" in the ingredient list.

Proportions for a packet of powdered culture come with the package. If you are using store-bought starter, plan on about two heaping tablespoons for each quart (1 L) of milk. After you have made your yogurt, you can save some as starter for subsequent batches. It will gradually give poorer results and eventually die out, at which time you can begin with fresh starter as before. If you are a goat purist, either use the packaged culture or look for an active-culture goat yogurt for your starter.

To make your yogurt, start by warming the milk to 100 to 110°F (38 to 43°C). Add the yogurt culture. If you prefer a firmer product, add ¼ cup (60 ml) of powdered milk to each 3 cups (710 ml) of goat milk (use dry goat milk if you can find it). Pour the warm milk plus culture into the cups of a commercial yogurt maker or your own homemade incubator (see sidebar below). The yogurt coagulates in 5 to 6 hours at a constant 100°F.

You can consider your yogurt ready when the curd has shrunk slightly from the sides of the container and it stays solid when the container is jiggled. Without the added milk solids, your yogurt will be much softer. The best way to test it is with a spoon and your taste buds.

Potential Yogurt-Making Problems

Before you get started, you need to be aware of a few problems you might encounter in making yogurt. First, don't use milk that contains antibiotics. Beside the fact that it is bad for the environment and dangerous for some humans to consume, even a small amount of antibiotics in the milk will kill the yogurt-making bacteria, and the milk won't "clabber," or curdle. It's these healthy lacto-bacilli, or lactic-acid bacteria, that coagulate the milk and create the yogurt taste. You can get rid of the "bad," illness-causing bacteria by pasteurizing your milk (if you choose) and sterilizing all equipment used in the process. If you sanitize with chlorine, rinse the equipment with fresh water before using it.

HOMEMADE INCUBATORS

A yogurt maker will automatically keep the milk at the proper temperature, but a preheated thermos wrapped in towels to help hold the heat in works just fine. Or pour the warm milk and culture into a casserole dish, set it in a warm oven, and leave it overnight with the heat off. A variation on this is to fill a stock pot with 100 to 110°F (38–43°C) water, set in the jars of cultured milk, and leave those in the prewarmed oven overnight. You can also make yogurt with a heating pad, or let it coagulate on the back of the old wood cookstove.

On a warm, sunny day, you can use sunlight — although sunshine isn't particularly good for goat-milk quality: pour the milk and culture into a glass-covered container, and set it in the sun. (A solar oven, however, gets much too hot unless you tend it carefully.)

Another option is to put your container of yogurt mixture into an empty ice chest with a container of hot water. It should stay warm enough for the cooking period as long as the lid is kept shut.

Temperatures that are too high (above 115°F [46°C]) will kill the lactic-acid bacteria, too. On the other hand, a temperature below 90°F (32°C) will cause the bacteria to work too slowly. They haven't died, so the yogurt might still set if given more time. Disturbing or moving the yogurt during incubation can also hamper clabbering and often results in puddles of whey as the protein chains are broken apart and the trapped fluid is released. You'll also get whey pools if the temperature is just a little too high or if you incubate the yogurt too long.

If your yogurt is too sour, you may have incubated it too long or used too much starter. An "off" taste is usually attributed to the milk, unclean utensils, or old or contaminated culture.

Greek Yogurt Cream Cheese

This is a delightful change from store-bought products. Pour yogurt into a butter muslin–lined colander. One quart (1 L) is a good measure to work with. Let the yogurt set all day or overnight. The amount of whey that drains off determines the final product. Use it at once or refrigerate, in a covered container, for up to 1 week.

Try this for a breakfast treat, as they do in the Middle East, or for a healthful snack: Form the yogurt cheese into small balls (if they're too soft, refrigerate them overnight). Roll the balls in olive oil, and sprinkle with paprika.

You can use yogurt cream cheese in unbaked pies, too. (Maybe we should have started out with more than a quart!) Combine 3 cups (710 ml) of yogurt cream cheese with 3 tablespoons of honey and 1 teaspoon of vanilla or grated orange peel, and stir until smooth. Pour into a 9-inch (23 cm) baked pie shell. Chill for 24 hours before serving. This pie

is delicious as is, but for a special treat you might want to add your favorite topping — perhaps strawberries, peaches, blueberries, or whatever is in season or suits your fancy.

Kefir

Kefir, a drink of Russian origin, is similar to yogurt, and it's incubated at room temperature (65 to 75°F [18 to 24°C]) for 24 hours. In addition to lactic-acid bacteria, it contains a lactose-fermenting yeast that gives it a unique fizzy character.

There are two forms of kefir. One is made with a yogurtlike culture, the other from small, curdlike particles called kefir grains. These are added to the milk to clabber it and can then be strained out and reused indefinitely if properly cared for. The cultures are available by mail order.

Koumiss

You'll need a slight sense of adventure to try koumiss or kumiss, also called milk beer. It originated in central Asia and was traditionally made from mare's milk, which is said to give it a moderate alcohol content (about 2.5%). But honestly, wouldn't you rather milk a goat? This is an Americanized version.

Thoroughly mix 1 quart (1 L) of fresh-from-the-goat unchilled milk with 4 teaspoons of sugar and 1 teaspoon of dry yeast. Let stand uncovered in a warm place for 10 hours. Pour it back and forth between two pitchers until it's foamy and smooth. Then store it in a jar with a tight-fitting lid in a warm place for another 24 hours. Chill. Stir before serving.

Butter

Butter is difficult to make from goat milk, because of its so-called natural homogenization (see chapter 2). If you are fortunate enough to own a cream separator, set the cream-adjusting screw to the finest setting for goat milk.

If you don't have a cream separator, you'll have a hard time getting enough cream to warrant cranking up the churn. However, it is possible to make butter from goat milk without a separator. One method, which will separate at least some of the cream, is to leave the milk in a flat pan with as much surface as possible exposed to the air. You'll be able to skim off some of the cream in about 24 hours. The cream will keep for a week in the refrigerator, so you can skim the cream from daily milkings for up to a week to accumulate enough to make butter. It helps if you are milking Nubians, which give milk with a high amount of butterfat.

The cream should be "ripened" for a week in the refrigerator or at room temperature for 1 day. Ripening in this case does two things: components in the milk start breaking down and natural bacteria start growing. Both create flavor. If you have been adding cream each day for a week, it probably doesn't need any more ripening.

Put the cream into the churn, but don't fill it more than half full. If you don't have a churn, use an electric mixer or a French whip, or just shake it in a canning jar with a tight lid, opening the lid once in a while to release the pressure.

The butter should begin coming together in about 20 minutes, in the form of pea-size grains. Drain off the buttermilk. This isn't cultured buttermilk like what you buy in the store, but it's drinkable and also good for cooking.

Now you have to work the butter. Using a spatula or wooden spoon, press the butter against the side of the bowl, and pour off the buttermilk that's pressed out.

Then wash the butter in cold water to get out any remaining traces of milk, which will cause the butter to spoil. Repeat the washing until the water comes off clear.

If you prefer salted butter, add salt to taste, and work it in.

You might be surprised to see that your goat butter is white. That's normal. If you want to make it yellow, use a special "Dandelion" butter coloring, available from goat-supply houses. Regular food coloring won't stick to the butterfat.

Cultured Buttermilk

The culture in buttermilk sold in stores is created by adding certain organisms to regular sweet milk and cultivating, or culturing, them.

The simplest way to culture buttermilk at home is to warm 1 quart (1 L) of fresh goat milk to 72°F (22°C), and then add 2 tablespoons of store-bought buttermilk (which, of course, will be from cow milk). Stir, and let sit at room temperature for 8 to 12 hours, or until it's as thick as you like it. Refrigerate. You can use 2 tablespoons of this batch to start the next batch. Buttermilk is also used as a culture in some cheeses, as in the Gouda recipe on page 231.

HOW ELSE CAN I USE GOAT MILK?

Whenever a recipe or formula calls for milk, you can use goat milk. Conversely, any recipe specifying goat milk will work just as well with cow milk. There is no need to search for special recipes. When you have fresh, wholesome, delicious goat milk, that's the only kind to consider using, whatever you're making.

17

RECIPES FOR GOAT PRODUCTS

People who have fresh, delicious milk from their own goats probably drink more milk than people who buy cow milk in stores. Even then, they frequently have a surplus and search for recipes in which milk is a major ingredient.

Making cheese and the other products covered in chapter 16 can use large quantities of milk, but this chapter includes more ideas that will help you keep smaller surpluses from going to waste. And remember, you don't need special recipes for goat milk. Use goat milk in any recipe calling for milk.

Similarly, you don't need special recipes for chevon. It can be used in any dish calling for beef, pork, or certainly lamb. It can replace beef or venison in sausage but is too lean to replace the pork.

Make full use of the goat products that are harvested from your backyard dairy. Some of the following are family favorites, while others are recipes readers have shared with *Countryside* magazine over the years.

YOGURT VARIATIONS

For a basic yogurt recipe, see page 233.

- **Bavarian cream.** For each 2 cups of yogurt, add cool but unset gelatin-dessert mix (your choice of flavors), made double-strength.

- **Sherbet on a stick.** Stir frozen juice concentrate (to taste) into yogurt, and spoon into small plastic cups. Insert a plastic spoon into the center, and freeze. To serve, unmold and use the spoons like handles.

- **Sour cream.** Spoon any amount of yogurt onto a clean cloth, draw up the corners, and hang it to drain for 3 hours, or until it's as firm as you want it. Use in any recipe calling for sour cream, but be warned that it breaks down under heat. To use it in beef Stroganoff or other cooked dishes, add cornstarch or flour as a stabilizer, and heat and stir gently.

Think outside the box when it comes to using the bounty of your goat barn. Besides food, you can also make soap (see p.250), like these fancy bars scraped with a serrated edge for decoration.

Milk Toast

When your goats are producing plenty of milk and your hens are providing more eggs than you need for your usual cooking and baking, try this unusual "waste-not-want-not" dish.

 2 quarts goat milk
 ½ teaspoon powdered cinnamon
 1 teaspoon salt
 1 tablespoon powdered sugar
 4 thin bread slices
 6 egg yolks

1. Boil the milk with the cinnamon, salt, and sugar.
2. Lay the bread in a deep dish, pour a little of the milk over it, and keep hot, but without burning it.
3. Beat the egg yolks, and add them to the reserved milk. Stir the mixture over low heat until thickened. Do not let it curdle.
4. Pour the egg-milk mixture over the bread, and serve.

For a metric conversion chart, see page 278.

Milk Soup

 ¼ cup diced onions
 1 gallon fresh goat milk
 2 cups egg noodles

Fry the onions in a small amount of butter until translucent (if the onions are not fried, they tend to curdle the milk). Bring the milk and onions to a simmer, and cook the noodles in milk until tender. Serve hot with cheddar cheese shredded over the top.

Devonshire Cream

In Devonshire, England, in the nineteenth century, this cream was used to make a very firm butter. But it was also considered a gourmet item in London, where it was eaten with fresh fruit. You can serve it with fresh berries or with warm scones and strawberry jam. Because goat milk butterfat doesn't like to separate, the yield isn't always enough to satisfy the whole family.

1. Let milk stand 24 hours in the winter; 12 hours when the weather is warm. (Here is a case where pasteurized milk would be safest because of the long setting time without refrigeration.)
2. Set the pan on the stove over very low heat, and heat the milk until it is quite hot. Don't let it boil; the longer the heating takes, the better. Do not stir. When it's ready, there will be thick undulations on the surface, and small rings will appear.
3. Set the pan in a cool place for 1 day.
4. Skim off the cream, and serve. The remaining skimmed milk will not be as thin as whey, but it will have a tang of its own and can be used in baking recipes in place of milk or other liquids.

Sweet Cheese

This is a delicious mild cheese.

 1 gallon milk
 1 pint buttermilk
 3 eggs, well beaten

1. Bring the milk to a boil.
2. Add the buttermilk and the eggs. Stir gently.
3. When the curd separates, drain and press.

Goat Milk Fudge

You'll need a candy thermometer for this fudge.

  2 cups sugar
2½ squares baking chocolate
  1 cup goat milk
 ¼ teaspoon salt
  1 cup nuts, chopped
  1 teaspoon vanilla
  1 tablespoon goat butter (see page 235)

1. Mix the sugar, chocolate, goat milk, and salt in a heavy saucepan. Cook over medium heat, stirring constantly.
2. Bring to 236°F (113°C) on a candy thermometer or to soft-ball stage (a few drops dribbled into cold water will form a soft ball).
3. Remove from heat. Add the nuts, vanilla, and goat butter. Beat until thick and creamy.
4. Pour into a buttered dish and cool. Cut into squares.

Goat Milk Pudding

You can also fill cream pies with this pudding.

2½ cups goat milk
 ½ cup sugar (use brown sugar for a butterscotch flavor)
     Pinch salt
  1 egg
  4 tablespoons cornstarch
  1 tablespoon goat butter (see page 235)
  1 teaspoon vanilla
  2 heaping tablespoons sweetened cocoa powder (for a chocolate pudding)

1. Mix 2 cups of the goat milk, the sugar, and the salt in a heavy saucepan. Heat slowly.
2. While the milk mixture is heating, beat the egg. Add to the milk mixture, and bring to the scalding point, stirring constantly.
3. Dissolve the cornstarch in the remaining ½ cup of milk, and add to the scalding milk, again stirring constantly. Stir until thickened, and remove from heat.

4. Add the goat butter and vanilla.
5. For flavored puddings, mix in the cocoa with the sugar before adding the milk, or substitute brown sugar for white for butterscotch flavor.

Vanilla Ice Cream I

This recipe is easy and is just right for one or two people.

2 eggs, separated
   Scant ½ cup powdered sugar
   A few drops vanilla extract
⅝ cup goat milk
⅝ cup heavy cream

1. Beat the egg yolks, sugar, and vanilla in a bowl.
2. Meanwhile, bring the milk to a boil. Pour it over the egg-sugar mixture, stirring constantly. Cool, and then refrigerate until quite cold.
3. Whisk the egg whites until stiff.
4. Lightly whip the cream.
5. Fold the egg whites and cream into the cold egg-sugar-milk mixture. Whisk well. Pour into a tabletop ice cream maker to finish, or follow steps 6 through 9.
6. Pour the mixture into shallow trays, and freeze until slushy.
7. Return the mixture to the bowl, and whisk again.
8. Pour back into trays, and freeze again.
9. When frozen, refrigerate for 30 minutes or so to soften it slightly.

Vanilla Ice Cream II

This recipe requires a little more work than the others. But if you enjoy the premium, high-priced, store-bought ice creams, it's worth the extra trouble.

4 eggs, lightly beaten
1½ cups sugar
½ teaspoon salt
2 cups goat milk
2 cups light cream
1 tablespoon vanilla powder*
4 cups heavy cream, well chilled

1. Combine the eggs, sugar, and salt in the top of a double boiler.
2. Whisk in the milk and light cream, and cook over simmering water, stirring constantly. When the mixture thickens slightly, remove it from the heat.
3. Add the vanilla powder, straining it through a large mesh sieve to remove any portion of the bean that is not finely ground.
4. Stir thoroughly, and refrigerate, preferably overnight but for at least several hours.
5. Just before you're ready to start cranking, remove the cold custard from the refrigerator, and blend in the cold heavy cream. Pour the mixture into the canister of the ice cream freezer, and crank.

*Note: Vanilla beans are expensive (and hard to find in some places), but they're a real treat. To make the powder, grind several dried vanilla beans in a spice mill. One 4-inch (10 cm) bean will make about 2 teaspoons of powder. The tiny specks of vanilla will show, but in ice cream, that's wonderful.

Honey Orange Sherbet

4   tablespoons lemon juice
½   cup orange juice
1   cup plus 2 tablespoons honey
1   quart goat milk

1. Mix the fruit juices and honey, and stir until liquid.
2. Slowly add the mixture to the milk. If it curdles, don't worry; the lumps will disappear while the mixture is freezing.
3. Process in a crank ice cream freezer.

Cooking Chevon

Milk is the most important reason for having goats, but they provide more than dairy products. Goat meat, or chevon, is excellent! Use it just like beef, lamb, or venison, or try some of the ethnic dishes that are popular now from the Caribbean, Asia, and Africa, where goat meat is a staple.

Chevon, like venison and buffalo meat, has a very low fat content. Rapid cooking at high temperatures, without added moisture, will result in a tough, dry, flavorless product. Meat of this type should be cooked slowly, and at low temperatures, with moist heat.

Chevon does have a distinctive flavor and aroma, quite unlike beef, pork, or venison. Many people value this difference and want recipes that preserve and enhance it. This might involve simply browning the meat in olive oil and roasting it with salt and pepper, so the natural flavor isn't masked. To discover why goat meat is so popular in Greece, add some oregano, garlic, and lemon juice. In other places, goat meat curry is preferred.

Some people don't appreciate the flavor, and if they have to eat chevon only because they have excess goats, they prefer recipes that hide or eliminate the taste. This is easily done with marinades or cooking it in any number of sauces, especially those containing tomato. Simmer it slowly with fresh or canned tomato chunks or sauce, fresh or dried celery leaves, garlic if you like that, oregano, cilantro, Worcestershire sauce, lemon juice, or anything else you fancy. Marinate the meat in cold black coffee or a cola beverage before roasting. Or grind the meat and mix it with ground beef or pork.

If you have cooking experience and creativity, you won't have any trouble preparing delicious meals with chevon. If you need some help, look for recipes in Mexican, Greek, and other ethnic cookbooks.

But also remember, you don't need any special recipes. In fact, in some households with fussy eaters, it might be wise to get your family accustomed to chevon by starting out with familiar dishes, especially stews, casseroles, or chili that involve a variety of ingredients and spices, slow cooking, and moist heat.

Chevon Chili

This recipe comes from Frank Pinkerton, "the goat man" of Langston University in Oklahoma. Frank does not recommend adding beans to this recipe. Serve pinto beans as a side dish. If you prefer a more traditional red chili, substitute 6 cups of stewed tomatoes and 2 cups of water for the final 8 cups of water.

2   cups onions, chopped
2   tablespoons olive oil
1   tablespoon ground oregano
2   tablespoons ground cumin
1   teaspoon garlic powder
1   tablespoon salt

TRY CHEVON IN ALL YOUR LAMB RECIPES

Chevon has a lamblike flavor, the cuts are similar, and like lamb, chevon is somewhat dry and quite lean. Because of that, any recipes for lamb are equally good with chevon — whether they call for roasting, baking, or grilling or for ground meat. Try ground chevon in homemade sausage.

3   pounds ground or cubed goat meat
½   cup chili powder, or to taste
½   cup flour
8   cups boiling water

1. Sauté the onions in the oil in a cast-iron pot, if you have one, or in a heavy Dutch oven or stockpot.
2. Add all spices except chili powder. Stir occasionally.
3. When the onions are almost clear, add the meat. Simmer until gray.
4. Add the chili powder and flour, and stir vigorously to thoroughly blend everything.
5. While stirring, add the boiling water, and bring the entire mixture to a boil.
6. Simmer for not more than 1 hour. Add other seasonings, such as cayenne or hot peppers, at this time, if desired.

Chevon Stew

Does anyone need a recipe for stew? Well, probably at first, but after that, if it's not instinctive, you aren't making stew. (Please forgive me, but next to goats, cooking from scratch, without recipes, is one of my greatest passions and pleasures. And for homesteaders like me, making something out of nothing and eating with the seasons is a virtue, a necessity, and a point of pride.) Here's how I make stew.

You know stew is on the menu when all the "fancier" cuts of meat have been used, the fresh summer vegetables are gone, and some of those in the root cellar are calling for attention. Or maybe it's just a cold wintry morning and you feel like simmering a kettle of stew on the woodstove all day. Take inventory, and gather your ingredients.

The meat is paramount, of course. It should be cubed, in bite-size squares. Sear the meat in olive oil, bacon grease, or butter, which tenderizes it. Even canola or vegetable oil will work, but being a peasant gourmet I really prefer olive oil or butter.

Add onions in any size or number that looks right to you. Toss in some homegrown garlic in whatever size. Powder works, if that's all you have.

Sprinkle some Worcestershire sauce over the meat, and stir it, then scatter in about a tablespoon of flour (be sure there's enough juice/oil at this point), and stir it well to coat the meat pieces. More flour, or less, won't hurt anything, but this step determines how thick the gravy will be. (You can thicken it some more later, if you want, by stirring in a flour-and-water mixture and letting it cook.)

The floured meat will be a thick, pasty glob. Add water to thin it, and stir it briskly to make a gravy.

Sprinkle this with crushed homegrown oregano leaves, basil, parsley, cilantro—whatever—but certainly celery leaves. You can use the stems, too, of course, but the leaves have the most flavor, and in our garden we get more leaves than stems.

Then add cut-up potatoes—as many as you think look good, in whatever shape or size appeals to you. They can be cubed, or you can use small whole ones. Carrots, definitely. And turnips, Jerusalem artichokes, rutabagas, parsnips, celeriac, corn, peas, green beans—whatever you have that wants eating and will fit in your stew pot.

I didn't mention green peppers (or red or yellow ones) or mushrooms (wild, from your inoculated logs, or store-bought), but by now you should be getting the idea.

Start this in the morning, and let it simmer on the woodstove all day, adding a little water as needed. Serve it with thick slices of fresh homemade bread or scones with homemade goat butter when the crew comes in from barn chores on a frosty winter evening, and they'll swear that chevon is as good as it gets.

Barbecue Sauce for Chevon Ribs

Chevon chops tend to be small, and I generally just leave them with the ribs. Barbecued ribs are popular at my house. Here's a tasty sauce.

- ½   cup onion, chopped
- 1   clove garlic, crushed
- 1   tablespoon fat or drippings
- ½   cup water
- 1   tablespoon vinegar
- 1   tablespoon Worcestershire sauce
- ¼   cup lemon juice
- 2   tablespoons brown sugar
- 1   cup chili sauce (or home-seasoned tomato sauce)
- ½   teaspoon salt
- ¼   teaspoon paprika
    Thyme, oregano, dry mustard, hot peppers, or pepper sauce (optional)

1. Sauté together the onion and garlic in the fat until tender.
2. Combine all the other ingredients and simmer for 20 minutes.
3. Pour the sauce over well-browned ribs and bake at 350°F (180°C) or grill over indirect medium heat until tender. This also works well in the slow cooker.

Jamaican Curried Goat

  3   pounds cubed chevon
  1   large onion, sliced
  5   cloves finely chopped garlic
  2   teaspoons salt
  1   teaspoon coarse black pepper
  1   teaspoon thyme leaves
  3   tablespoons olive oil
  1   teaspoon sugar
  5   scallions, chopped
2½   teaspoons curry powder
  2   potatoes (optional)
  1   teaspoon cornstarch (optional)

1. Place the chevon, onion, garlic, salt, pepper, and thyme in a sealed container, and shake well to coat the meat. Allow to marinate in the refrigerator for 2 hours.
2. Heat the olive oil in a stew pot over medium heat. Add the sugar, and warm until the sugar is brown.
3. Add the marinated meat and remaining spices.
4. Add the scallions, curry powder, and ¼ cup warm water; stir thoroughly. Cover the pot, and allow to simmer slowly for 30 to 45 minutes, stirring occasionally until the chevon is tender.
5. Some people like to add a couple of peeled and cubed (½ inch) potatoes for the last 15 minutes, but the curry is also tasty without the extra starch and served over rice. If the mixture needs thickening, use ½ to 1 teaspoon cornstarch in a little water; mix in briskly, and cook until thickened.

Broiler/Grill Chevon Rounds

1. Make a burger-size patty of ground chevon. Season with salt, pepper, and garlic to taste.
2. Wrap a slice of thick bacon around the circumference of the patty, and secure it with a wooden toothpick.
3. Broil or grill the patty on high heat to sear the surfaces. Turn the heat down, and broil or grill on each side about 4 minutes or until cooked through. For a flavor burst, brush on some of the same barbecue sauce used for chevon ribs (see page 245) for the last minute of broiling on each side.

Greek Goat and Pasta

A friend gave me this recipe; it was a favorite of one of his Greek relatives. The ingredients were sketchy, including how big a goat to use, but he always took home a 35- to 40-pound kid. When dressed, that would leave 17 to 20 pounds of meat, which would just about fit in a large electric roaster or an oven pan.

  6   garlic cloves
17–20   pounds chevon
      Salt
      Freshly cracked black pepper
  2   tablespoons Greek oregano
  1   teaspoon thyme
  2   tablespoons fresh mint, chopped
      Juice of 1 lemon
      Tomato sauce of your choice
      Pasta of your choice, enough for each guest
      Grated Romano or Parmesan cheese, for serving

1. Slice about a half dozen garlic cloves lengthwise, puncture the meat, and push the garlic pieces in the holes.
2. Mix the salt, pepper, oregano, thyme, and mint flakes, and rub the mix liberally on the meat.
3. Put the meat in a roaster on a rack, and bake, covered, 15 to 20 minutes on each side at 425°F (220°C).
4. Reduce the heat to 375°F (190°C), and cover the bottom of the pan with hot water.
5. Mix 1 cup of hot water and the juice of 1 lemon to pour over the meat. Roast until done (1 to 2 hours, depending on the size of the piece of meat), basting with juices from the bottom of the pan every 15 minutes. Add more water if needed, and turn the meat over in the pan about halfway through. Leave uncovered (but keep basting) for the last 20 minutes of roasting.
6. When the meat is done, remove the juices and strain them into a pot. Add enough canned or fresh tomatoes to make a sauce and heat. The quantity of tomato will depend on how many people you are feeding and how saucy you want your pasta side dish.
7. Parboil the pasta in a separate pot, drain, and toss with the meat juice/tomato mixture. Serve the meat with the pasta, topped with grated Romano or Parmesan cheese.

Marinating Chevon

Any meat can be marinated. The essential marinade ingredient is an acidic liquid, such as wine, vinegar, lemon juice, tomato juice, or cold coffee. The acid tenderizes the meat and enhances the flavor.

There are, of course, many marinades. If you're in a hurry, just splash some red wine and water on the meat in a dish, and sprinkle on whatever herbs and spices you like (garlic and oregano are the standards at our house). Try the following suggestion if you have time for something a little more elaborate.

Marinated Chevon Steak

This recipe is for about 3 pounds of chevon steak.

3	pounds chevon steak
	Cider vinegar and water (half and half) to cover the meat
¼	cup honey
1	large onion, sliced
2	teaspoons salt
¼	teaspoon paprika
4	cloves, whole
2	bay leaves
1	teaspoon oregano
½	teaspoon dry mustard
1	cup thick sour cream

1. Arrange the steak in a shallow glass dish. Cover with the water and vinegar.
2. Add everything else except the sour cream.
3. Marinate in the refrigerator for 2 days.
4. Remove the meat, pat it dry, and dredge it in flour and butter. Brown the meat.
5. Add the marinade, and simmer gently until tender.
6. Stir in the sour cream, and serve.

Making Sausage with Chevon

Sausage making is another one of those projects that can convert a backyard dairy into a homestead or become a fascinating hobby in itself. As with cheeses, there are hundreds of sausage recipes. While they all have the same starting point and certain steps in common, chevon sausage recipes can be very simple or very complex; and they can easily be made in your kitchen with very little special equipment.

D. L. Salsbury, a veterinarian and an avid sausage maker who has shared many recipes with *Countryside* readers, likes to point out that while some people think sausage making has to be complicated, if you have ever made meatloaf, you have made sausage. It's the same basic process.

Any butchering project will end up with bits and pieces of meat that might not be suitable even for stews; these are ideal for sausage. Older cull animals might best be utilized by using the entire carcass for sausage. But any meat, even the best cuts, can be included in sausage.

When making sausage with chevon, you must add at least some pork fat because chevon is lean and dry. It's the same with venison, so venison sausage recipes, which are quite common, work very well with chevon. You might even be able to find premixed spices and curing agents for venison sausage in your local supermarket — at least during hunting season.

For starters, try a simple recipe, one that doesn't require casings and stuffing, smoking, or aging. Here's an example.

Chevon Sausage

You can alter this recipe easily to suit your own tastes. You might want to add garlic, onions, your favorite herbs, or lemon juice. You can eliminate the potato if you want a meatier, less meatloaf-type sausage. (The egg is essential; it acts as a binder and replaces pork fat in the recipe.) Try

adding finely diced green or red peppers or hot peppers, if you enjoy those. Be creative! Have fun!

MAKES 4 SERVINGS

- 1 pound chevon, ground
- 1 medium potato, boiled, peeled, and mashed
- ½ cup spinach, chard, or dandelion greens, parboiled, drained, pressed dry, and pureed
- 1 egg
- 2 tablespoons Parmesan cheese, grated
  Salt and pepper, to taste
- ¼ cup bread crumbs, plus extra as needed
- ¼ cup flour
- 4 tablespoons butter
- 1 tablespoon olive oil
- ½ cup white wine
- 2 cups beef stock

1. With clean hands (no rings), mix the meat, potato, greens, egg, cheese, salt, and pepper in a bowl.
2. Add just enough bread crumbs, if necessary, to make the mixture hold together when formed into a ball. Mash it well, squeezing it between your fingers.
3. Then shape the meat mixture into a loaf or cylinder. Combine the flour with the ¼ cup of bread crumbs, and coat the meat with this mixture.
4. Brown the meat on all sides in the butter and oil.
5. Add the wine.
6. When the wine has almost evaporated, add the stock, cover, and simmer over low heat, turning occasionally, for 1 hour, or until the inside is done.
7. Remove the meat from the pan, and boil down the liquid until it's thickened.
8. Slice the loaf, and pour the juice over the slices to serve.

Saucisses De Chevre

In any language and in any setting, goat products like these sausages are irresistible.

Soapmaking with Goat Milk

Like cheesemaking and sausage making, soap-making can become a pleasurable hobby. You won't use up much milk making soap, although I do know someone who got into goats as the result of obsessive soap making.

Goat milk is used in soap recipes because the natural fats add moisture to the skin. Goat-milk soap also has the mysterious property of removing buck smell from hands.

Goat-Milk Soap

 3 pounds goat lard (or other animal fat), rendered and clarified
17 ounces olive oil
12 ounces safflower oil
 8 ounces canola oil
 3 pounds (6 cups) pasteurized and frozen goat milk
12 ounces lye (sodium hydroxide)
 2 cups oatmeal, finely ground
 1 ounce borax (sodium borate)
½ ounce glycerin

1. Melt the lard in an 8-quart pot over low heat, add the liquid oils, and heat to 110°F (43°C) and no more.

2. Put the frozen milk in a 4-quart pot. With safety wear in place, very slowly pour the lye over the icy milk, stirring with a sturdy plastic or stainless steel spoon until the ice is dissolved and the powder thoroughly mixed. The icy milk will help control the extreme heat that normally builds up when lye is mixed with water. Beware of fumes. Set the lye mixture aside until it cools to 85°F (30°C).

3. While stirring, very carefully and slowly pour the lye/milk mixture into the warm fats (not the other way around!). It should take at least 15 minutes. An electric hand mixer makes the job easier, but be careful not to splash the mixture because it will burn.

4. Add the oatmeal, borax, and glycerin, and continue to stir or mix until the ingredients stay joined and the spoon or mixer leaves a silvery trail through the thickening mass.

5. Pour the mixture into molds (plastic-foam drinking cups make fine molds) or into a shallow wooden box especially made for square soap molds.

SOAP SAFETY

While it is a great accomplishment to make your own soap, the process is inherently risky and should not be attempted with children nearby. Eye protection and rubber gloves should be worn when working with the lye solution and again when handling raw, unripened soap. Use dedicated utensils for making soap, and don't return them to the pantry for making food. Work in a well-ventilated room, and don't breathe the fumes. Use stainless steel or enamel cookware, plastic, or glass measuring instruments, and heavy plastic or stainless steel stirring spoons. Protect your work area with newspaper or plastic. Work in weights of ingredients rather than volume for more accuracy.

6. Cover the molds with a thick towel for 24 hours so they don't cool off too quickly. If the soap is in a box, cut bars after 24 hours but leave them in the mold for another 24 hours.

7. Remove the soaps from the molds, and let them ripen 3 or more weeks in a cool, dry place. Long ripening makes a better bar that will last longer.

OTHER IDEAS

Several books are available for the home soap maker, including a couple that are especially for milk-based soaps, such as *Milk-Based Soaps* (Storey Publishing, 1997). Don't forget to try some of the other personal care items, like goat-milk hand cream and milk bath.

What's Next?

We haven't mentioned making useful and beautiful leather and fur from goat hides (tanned kid skins are very much like fur). You can make candles from goat tallow. When you become accustomed to dining on chevon and making sausage, there will be no such thing as "unwanted" kids. When you get really good at making cheese, yogurt, butter, and ice cream, and you use fresh delicious goat milk in custards, baking, and other cooking, you might find that you don't have enough milk left to drink, which would be a tragedy.

The solution? Simple! Go back to the beginning of this book. Then get more goats!

WHERE MILK COMES FROM

What makes a goat "let down" her milk? Why does she sometimes "hold back" milk? What accounts for the lactation curve, and what makes a goat dry off? In brief, where does milk come from? As a goat owner, you're almost certain to ponder these questions while milking, sooner or later. Here are some of the answers.

Like all mammals, female goats produce milk for the purpose of feeding their young. Thus, the doe must be bred and give birth before she will lactate (with certain abnormal exceptions), and lactation stops naturally as kids are weaned. This period of milk production can be extended somewhat but not indefinitely. Eventually, the doe must be bred before she will start producing milk again.

Activity in the Udder

The goat has two mammary glands, which collectively are called the udder. The udder is not an organ as such. It is actually part of the body's largest organ — the skin.

Probably less than 50 percent of the milk an animal produces can be contained in the natural storage area of the udder. The balance is accommodated only by the stretching of the udder. In some cases, this can cause the ligaments that suspend the udder to become permanently lengthened, causing the udder to "break away" from the body, resulting in what is called a pendulous udder.

A good udder is capacious, has a relatively level floor, and has a broad and tight attachment to the body. It has plenty of glandular tissue but a minimum of connective and almost no fatty tissue. After milking, the normal high-quality udder feels soft and pliable, with no lumps or knots that would indicate scar tissue resulting from injury or disease.

The udder is divided into right and left halves by a heavy internal membrane. The milk produced in each half can be removed only from the teat of that half.

The Teat, Lobes, and Alveoli

Why are some goats hard to milk and others "leakers"? Let's take a closer look at the mammary system, starting with the teat, and working back to the origin of the milk itself.

At the tip of the teat is an orifice, actually a short canal called the "streak canal." The streak canal is surrounded by sphincter muscles that act like a flow restrictor on a garden hose. These muscles prevent the milk from flowing out, and their strength determines how hard or easy it is to milk the animal. Goats with very strong sphincters might be hard to milk, while those with weak sphincters can actually drip milk (we'll see that these sphincters are affected by hormones resulting from premilking routines and other factors, and why it's important to avoid changing or upsetting those routines, among other things).

Following the path of the milk back to its origins, we see where the teat widens into its cistern, the final temporary storage area before it is removed by milking or suckling. Farther up is a larger storage area called the gland cistern, which is in the udder; and beyond that are large ducts that branch through all parts of the udder to collect and transport the milk toward the teat.

Each duct drains a single lobe, which further branches into lobules, where milk production actually takes place.

These lobules are composed of many tiny, hollow spheres called alveoli, which are said to resemble bunches of grapes — very tiny grapes. One cubic centimeter contains about 60,000 alveoli. Milk is secreted in the cells of each alveolus.

The alveolus is surrounded by muscle fibers that balloon out as milk is secreted. When the animal is properly stimulated to let down her milk, these microscopic muscles contract, forcing out the milk. Inadequate or improper stimulation can therefore result in "hard milking."

Blood Supply

Blood supply to the udder is important in the making of milk. For each unit of milk secreted, anywhere from 400 to 500 times as much blood passes through the udder!

Blood enters through the base of each half of the udder through the pudic and then mammary arteries. Those arteries branch off into smaller and smaller arteries, vessels, and capillaries for both halves of the udder. The system of smaller blood vessels may be more important than the major arteries. In research, when an artery was experimentally severed, milk production dropped to almost zero for a few days, but smaller subsidiary arteries increased in size, and production gradually returned to normal. Interestingly, the blood vessels that can be seen near the surface of the udder and the belly are actually veins. They return depleted blood to the heart. One of these is the subcutaneous abdominal vein, often called the milk vein. Many milkers believe the size and prominence of the milk vein is an indication of milking ability.

Milk is made largely from constituents of blood. The tiny network of capillaries that surround each alveolus carries blood to the base of the cells that line the alveolus. Milk is partially made of materials in the blood that transfer through the thin walls of the capillaries into the milk-making cells. Some of these constituents are used as is, but others are synthesized into new elements through cellular metabolism. Water, the largest component of milk (87 percent by weight), is filtered from the blood, as are milk's vitamins and minerals. The lactose in milk (its principal carbohydrate) is synthesized from blood glucose.

On the other hand, about 75 percent of the fat in milk is synthesized in the mammary gland. Most of this comes from acetate, which explains why animals on high-grain and low-forage diets often produce milk with a lower fat content. This kind of diet results in reduced production of acetate in the rumen.

Secretion of the Milk

Now it's milking time. The cells of the alveoli have been busy making milk by filtering blood and synthesizing other constituents; these tiny cells have lengthened as the milk accumulated. When filled, the cells ruptured, pouring their contents into the lumen, the cavity of the alveolus. This caused the mammary glands to become saturated with milk, like a sponge.

Incidentally, all this activity is most rapid immediately after milking but gradually slows as the udder becomes full in preparation for the next milking. As milk is secreted in the cells and collected, increasing pressure in the mammary system slows the secretion-discharge cycles. Each hour after milking, milk production decreases by 90 to 95 percent. At a certain point (technically, at a pressure of 30 to 40 millimeters in mercury, which is roughly equal to capillary pressure and about one-fourth of systemic blood pressure), milk secretion is reduced appreciably or stops entirely. If the animal isn't milked (or if its kid dies in the wild), the milk starts to be reabsorbed into the bloodstream.

Also of practical interest is the fact that, when the udder is full, the secreting cells are unable to rupture because of the pressure. Therefore, only that part of the milk that can pass through the semipermeable cell membrane can be discharged. The milk fat is not discharged. This lower-fat milk dilutes the milk previously secreted. That's why when the interval between milkings lengthens, the fat content of the milk decreases.

This also explains several other important phenomena, such as why milking three times a day results in more milk than milking twice a day. Each time the udder is emptied, it is followed by a period of fast production. If that hyperproduction happens three times a day, more milk is produced overall than if it happens two times a day. This is also the reason the last milk out of the udder is higher in fat than the first milk drawn; the cells with accumulated milk fat are able to discharge the fat globules when udder pressure has been reduced. And in addition, it helps explain why the milk-fat percentage is usually higher with lower-producing animals or those whose production is declining in late lactation. For both

animals, there is less pressure on the udder to prevent discharge of the fat globules.

Now the stage is set. The goat is almost ready for you to start milking. Almost, but not quite! Coaxing milk from the udder involves more than just squeezing teats. What are required now are the only stimulators of lactation: hormones.

Hormones

Hormones are closely related to milk production in many ways, one of the most obvious being the development of the udder itself. The udder is, after all, a secondary sex characteristic. Its very existence and function are the results of hormonal activity.

Six hormones are important in the intensity of lactation.

1. **Prolactin** is secreted by the anterior pituitary gland and in mammals stimulates the initiation of lactation. It also increases the activity of the enzymes that are essential to the work of the epithelial cells (in the alveoli), which convert blood constituents to milk.

2. **Thyroxine** is secreted by the thyroid gland. Cows experimentally deprived of this product of thyroxine have gone down in milk production by as much as 75 percent. Other research has shown that thyroxine secretion increases in the fall and winter and decreases in spring and summer. This is said to partially explain why milk production decreases in hot weather.

3. **Somatotropin,** secreted by the anterior pituitary gland (known as bovine growth hormone, or BGH, in cows), regulates

growth in young animals but also influences milk secretion by increasing the availability of blood amino acids, fats, and sugars for use by the mammary gland cells in milk synthesis.

4. **Parathyroid hormone,** secreted by the parathyroid gland, regulates blood levels of calcium and phosphorus, which are major constituents of milk. It also plays a role in milk fever (see chapter 8): When an animal freshens and starts producing milk, the mammary glands rapidly withdraw calcium and phosphorus from the blood. Without proper nutrition and without parathyroid hormone, the animal can develop milk fever. Low vitamin D tends to depress calcium levels, so it's typical to feed more of the vitamin before freshening.

5. **Adrenal hormones,** products of the adrenal glands, work both ways — small amounts are essential to milk production, but larger amounts will depress it. This is why it's important not to startle or frighten dairy animals. When a doe is disturbed, adrenaline is secreted to overcome the stress of the moment, but it decreases milk secretion.

6. **Oxytocin** is secreted by the hypothalamus and works with prolactin (it also induces expulsion of the egg in the hen and is used to induce active labor in women or to cause contraction of the uterus after delivery of the placenta).

Stimulation of Hormones

A goat doesn't voluntarily "hold back" her milk. But she does have to be properly stimulated. When she is, milk is suddenly expelled from the alveoli into the large ducts and udder cisterns. Then, and only then, can it be removed.

The natural stimulus is nursing. However, manual massage of the teats and udder, performed while washing them, has the same effect. In addition, sights, sounds, and smells have an effect on milk letdown. Your arrival at the barn, turning on the lights, feeding: all the routines of milking are signals for the milk to start pumping. A change in those routines is one of the reasons many goats will produce less when moved to a new home.

When the letdown occurs, oxytocin is poured into the bloodstream, reaching the udder in 30 to 40 seconds. This causes the cells to contract, squeezing milk from the alveoli. Milk pressure in the cistern is almost doubled. However, this lasts for only 5 or 6 minutes (this backs up the belief that to get the most milk you must milk fast).

Oxytocin can be neutralized by adrenaline. This hormone increases blood pressure, heart rate, and cardiac output. It also causes the tiny arteries and capillaries of the udder to constrict and thus prevents oxytocin in the blood from reaching its destination. And adrenaline remains in the blood longer than oxytocin.

Adrenaline, of course, is released when the animal is in pain, frightened, irritated, startled by a loud noise, or otherwise bothered. Small wonder that the first-time milker who pokes, frightens, irritates, startles, and probably embarrasses the goat gets so little milk!

Drying Off, or Involution

You never get the "last drop" of milk from an udder. Normally, 10 to 25 percent remains even after stripping.

But if the available milk isn't removed from the udder, the milk already there will be reabsorbed and the cells will quit producing more. Even just incomplete milking causes pressure to build more rapidly, and less milk is secreted between milkings. Eventually, the secretion is impaired, and the animal dries off.

A goat can be dried off by milking on alternate days or by stopping milking altogether. Quitting cold turkey is preferred.

Other Factors Affecting Lactation

As we've seen, hormones play a major role in milk production. But they aren't the only factors. There are many others.

1. **Genetics.** Genetics affect milk production. If the animal hasn't inherited the potential to produce milk, including the capability to produce the needed enzymes, she won't produce milk.

2. **Secreting tissue.** The animal must have enough secreting tissue. A small udder has proportionately less of this tissue, and it also has less capacity for storing the milk secreted (this doesn't necessarily mean a large udder is better, but an udder should have capacity).

3. **Stage of lactation.** Milk production generally peaks within 1 to 2 months of freshening, then drops off. The rate of this drop-off has a marked effect on annual production. Persistent milkers might drop 2 or 3 percent a month; others drop off much faster. Easy milkers, those that milk out more rapidly, are usually more persistent.

4. **Frequency of milking.** As we've seen, frequent milking tends to lengthen lactations.

5. **Stage of pregnancy.** Females in later stages of pregnancy generally drop off in milk production quite rapidly, because nutrients are diverted from the mammary gland to the uterus, for growth and maintenance of the fetus.

6. **Age.** Milk production increases with age, up to a point, and then drops off with advancing years.

7. **Estrous cycle.** Milk production drops temporarily when an animal is in heat.

8. **Health.** As might be expected, any disease can reduce milk production. Diseases can slow the circulation of blood to the udder, and that again, for reasons already given, affects milk secretion.

9. **Feed and nutrition.** In the hierarchy that nature has wisely established, a body's first responsibility is survival. Maintenance comes before milk production.

10. **Environmental temperature.** High temperatures decrease milk production. This has several explanations, including depressed appetites, reduced thyroid secretion, and others. Optimum temperatures seem to be between 50 and 80°F (10 and 27°C).

11. **Milking routine.** If you don't properly stimulate your goat, you won't get all the milk possible, and that in turn will decrease production at future milkings.

Milking isn't just a chore; it's participating in a marvel of nature. And lucky people who have goats get to do it twice a day!

SOMATIC CELL COUNTS

As beginning goat milkers and goat-milk drinkers, our first interests are likely to concern quantity, flavor, and basic food safety (such as animal health factors associated with milk and sanitation procedures). You could milk goats for a long time without even hearing about somatic cell counts (SCCs). Sooner or later you will — and you'll probably wonder what all the fuss is about. Here's a simplified explanation.

Somatic cells are white blood cells that are routinely sloughed off into the milk. Excessive amounts can indicate mastitis, which is caused by infectious bacteria in the milk. "Counting" the number of these cells in a milliliter of milk is the basis for mastitis tests. For commercial dairies, the SCC must be below certain levels. So far, so good. But goats are not little cows, and the source of the cells that are being counted can be very different in goats, and very often the number of these cells does not indicate mastitis.

Just to give you an idea of the numbers we are talking about, a commercial Grade A dairy must maintain an SCC of 750,000 cells per milliliter or below for cows. For many years, the level for goats remained at 1 million, but research and sensible minds prevailed in 2009, and the level for goats was raised to 1,500,000, although some states have been slow to change the language in their statutes. An average cow dairy has an SCC of around 300,000 and well-managed herds can come in at under 100,000. Not so with goats, where even a well-managed herd is likely to have an SCC of 800,000 or higher at certain times of the year.

First, let's look at mastitis. When there is an infection in the udder, white blood cells are called into the udder to engulf and destroy the bacteria. They "leak" through the walls of the blood vessels that serve the udder and into the secretory cells. The white blood cells are then secreted into the udder and often carry some of the cells — the somatic cells — with them. The process does damage to the mammary and, if left untreated, will cause permanent damage, reducing milk production. A report in the *Journal of Dairy Science* found that infected udder halves in test groups produced 1.5 pounds of milk compared to 2.2 pounds in the uninfected half. The infected half had an SCC of 1,700,000, and the SCC in the uninfected half was 400,000. Mastitis will show up as clotted and stringy milk in the strip cup and as a high SCC if tested.

On the other hand, improper identification can produce a high somatic cell count. With goats, when secretory cells in the alveoli rupture to release milk, small pieces of the cell are sloughed off at the same time. Some mechanical forms of counting SCCs can't tell the difference between a partial cell and a whole cell, and the counts can sometimes look much higher than they actually are. Cell counts using Pyronin Y-methyl green stain are much more accurate, but they depend on human accuracy to count thousands of little dots on a slide. One Wisconsin milker sent samples from the same animal on the same milking and received results with a 300,000 SCC difference.

For no reason that has been pinned down by science, goats also tend to toss off somatic cells when they are stressed, when routines are changed, or just when they are feeling amorous. Commercial dairies routinely have much higher SCCs in the fall and winter when does are in estrus. That time of year is even more problematic, because it is late in the natural lactation curve, when milk production is dropping and there is less milk to dilute the cell numbers. Commercial producers try for out-of-season breedings partly because they get a premium for winter milk and partly so there is more milk in the bulk tank in the fall to keep SCC numbers within limits.

In cows, high SCCs also decrease the amount of cheese a given quantity of milk yields. Not so with goats, as long as the udder is healthy. If the goat's high SCC is caused by mastitis, its milk will also produce a poor cheese yield with long coagulation times.

THE COMPOSITION OF MILK

"Goat milk is richer than cow milk, isn't it?" If we assume "richer" means higher in fat, the answer is "sometimes."

There is a great deal of variation in the composition of milk, not only among species but also breeds, families, and even the same individuals at different ages and stages of lactation or on different feeds.

If you look at one of those charts giving the "average" composition of milk (and seldom do two of them agree), here's what you might find:

Similar charts for various breeds of cows show greater differences than those indicated between cows and goats. The same is true for goats. While averages show that Nubians produce milk that's richer in fat than Saanens, these, too, are only averages. Some Saanens produce milk that's richer than the milk of some Nubians.

Some of the reasons are explained in Where Milk Comes From, page 252. Further, milk composition is highly heritable. Animals of the same bloodlines are likely to have the same milk characteristics (given the same health, feed, age, state of lactation, and other factors). Also, the milk of individual animals changes drastically at different periods of time: consider colostrum.

While colostrum presents an extreme example, changes in milk composition do occur throughout lactation, with milk during the first 2 months after freshening generally being from 0.5 to 1.5 percent lower in fat than milk from the same animal during the last 2 months of lactation.

Protein, fat, and solids not fat (SNF) generally decline with age, with SNF declining even more than fat. Butterfat tests normally rise in the fall and drop in spring.

Other factors that can cause minor changes in the composition of milk include temperature of the environment (temperatures above 70°F [21°C] and below 30°F [–1°C] increase the fat content) and exercise (slight exercise slightly increases fat content, while more strenuous exercise decreases fat and total output).

AVERAGE COMPOSITION OF MILK IN MAMMALS

Mammal	Fat	Protein	Lactose	Minerals	Total Solids
Human	3.7	1.6	7.0	0.2	12.5
Cow	4.0	3.3	5.0	0.7	13.0
Goat	4.1	3.7	4.2	0.8	12.9

COMPARATIVE COMPOSITION OF COLOSTRUM AND NORMAL MILK IN SEVERAL SPECIES

Colostrum is rich in proteins, in large part from the globulin contents that contain the antibodies that protect newborns, but the relative value can be very different among mammals. In a 2009 study in the *Journal of Mammalogy*, some of the key components of colostrum were compared to the same components in "mature milk" of the same species.

	Colostrum			Mature Milk		
	Protein (g/kg)	Fat (g/kg)	Carbohydrate (lactose) (g/kg)	Protein (g/kg)	Fat (g/kg)	Carbohydrate (lactose) (g/kg)
Goat	80	90	25	30	34	45
Sheep	130	124	34	58	60	43
Cow	130	36	31	33	38	50
Human	22.9	29.5	57	10.6	45.4	71

GIVING INJECTIONS

How do you give a goat an injection?

The best way to learn is by watching someone else. Most veterinarians will show you how to do something as basic as this; they have more important things to do with their time and education than giving shots. (Veterinarians don't "give shots," of course, or even injections; they give parenteral medication. This game of semantics just means the medication isn't administered orally.)

Injections can be intramuscular (IM), subcutaneous (SC or SubQ), intraperitoneal (IP), and intravenous (IV). These names refer to where the needle goes in. Intramuscular means within the muscle, or meat. Generally, this means into a rear leg or in the neck muscle. Subcutaneous means under the skin. Intraperitoneal means within the peritoneum, or the space between the intestines and other parts of the lower gut and the internal organs such as the liver. Intravenous means directly into a vein. The last two are best left to a veterinarian.

Different medications require different types of injections. Read all labels carefully—and not only to learn how the medication should be administered. Note also any safety precautions, how the medicine should be stored, shelf life, withdrawal time, and any other information provided.

Very few medications have been approved for goats. Any off-label or extra-label use of medications can only legally be administered by a veterinarian and under a strict veterinarian-client-patient relationship.

There are three commonly used types of syringes: all glass, glass and metal, and plastic. Most farms use disposable plastic syringes, as these don't require sterilization equipment. (*Note:* Equipment can be sterilized in a domestic pressure cooker or canner by placing it in steam under pressure for 30 minutes. Remove metal plungers so they do not expand and break the syringe. Letting syringes stand in a pan of boiling water for 30 minutes is another method, but it's not totally reliable as a sterilization method.) Disposable syringes come in sterile packages, are used once, and discarded.

There are also reusable needles, which require sterilization, and disposable needles.

These come in various lengths and gauges. For goats, 1-inch (2.5 cm) needles in 16-, 18-, and 20-gauge sizes are commonly used. The larger the gauge, the smaller the needle. The choice usually depends on how viscous the medication is that is being given, although it is also largely a matter of personal preference.

Syringes can be disposed of in the public trash, but proper needle disposal is a matter of public safety. Save used needles in a sturdy plastic container with a screw-on lid. When it is full, take it to your local hospital or veterinarian, where the needles will be disposed of in a "sharps" container specifically designed for the purpose.

There are several ways to fill the syringe. Here is one method.

1. Wipe the rubber bottle cap with alcohol or disinfectant.

2. Place a sterile needle on a sterile syringe.

3. Invert the medicine bottle, and insert the needle through the rubber cap.

4. Draw out a little more medicine than will be needed.

5. With the needle still inserted in the inverted bottle, tap the side of the needle to dislodge any trapped air bubbles and float them to the top of the syringe.

6. Depress the plunger to remove any air from the syringe and needle and expel any extra medication to get to the proper dosage.

7. Remove the needle from the bottle, and you're ready to go.

Some sources recommend that you fill the syringe with air and blow it into the bottle before withdrawing fluid. It may be necessary when a big bottle is nearly empty and a vacuum has formed from constantly removing medication. However, it is also blowing airborne bacteria into the bottle and possibly contaminating the contents.

Never reinsert a needle in a bottle of medicine after it has been used on an animal.

It's not possible to sterilize the skin of a live animal, but it should be as clean as possible. A wide variety of skin disinfectants is available, such as a 70 percent solution of ethyl alcohol in water and iodine, including tincture of iodine and Lugol's iodine.

Most vaccines and antitoxins are given subcutaneously. The needle is inserted through the skin, and the substance is deposited beneath the skin.

1. Grasp the skin between the thumb and forefinger, and raise it into a "tent" or bubble.

2. Insert the needle at the base of this tent, parallel to the muscle wall. Release the skin when the needle is in.

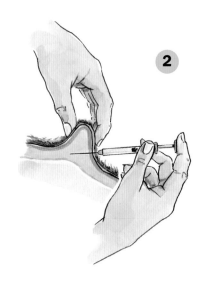

3. Push the plunger of the syringe to force its contents under the skin.

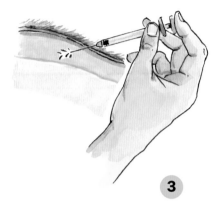

3

4. Withdraw the needle and syringe.

5. Gentle massage of the area has been shown to increase the rate at which the medicine is absorbed into the tissue.

The best spot for antibiotics and vaccinations is a hand's breadth back from the point of the shoulder in the fleshy tissue just ahead of the rib cage. There are no major vessels or nerves to hit accidentally. Hormones like Lutalyse and prostaglandin must be given deeply into the muscle. The lower thigh is recommended.

Follow the manufacturer's instructions exactly when using and storing vaccines and medications.

IMMUNIZATIONS

Every region has different health problems that can be prevented by immunizations. The person best qualified to help you set up and administer a health maintenance program is your veterinarian.

Contacting a veterinarian about vaccinations has several side benefits: it establishes a valid veterinarian-client-patient relationship; the doctor will know you have goats and that you care enough about them to seek professional guidance before some dire emergency arises; and a veterinarian is more likely to proffer valuable advice or assistance on other matters on a routine, nonemergency call than in a life-or-death situation.

Chances are your veterinarian will recommend vaccinations for tetanus and enterotoxemia, which come together in a single vaccine called CDT (*Clostridium perfringens* types C and D, plus tetanus toxoid). Two different studies of commercially manufactured vaccines compared antibody levels in herds of goats and sheep and showed that the goats did not stay protected as long as the sheep did. One of the studies found that most of the goats that were given a vaccination and a 28-day booster had antibody levels below the protective threshold by 98 days after the initial dose. The current recommendation is that vaccinations be repeated on a 3- or 4-month schedule for optimum coverage, and especially where enterotoxemia is a problem. For does, one of those vaccinations should be 45 to 60 days before kidding. Kids should be vaccinated at 6 weeks or at weaning and given a booster according to the bottle label.

If your part of the country is selenium deficient, an injectable dose of selenium may be recommended for the doe 2 to 4 weeks before kidding and for the kid 2 to 4 weeks after it hits the ground.

The most prevalent goat ailment is probably caseous lymphadenitis (CL), and the most effective vaccine must be created specifically for your farm and your goats, which is quite expensive. A general CL vaccine developed in Australia has shown fairly good results and has been released in 11 other countries but not the United States. A Colorado company has a sheep CL vaccine, but when this has been used in goats, there have been anecdotal reports of a severe drop in milk, lameness, fever, and depression for a few days after animals received the vaccination.

The United States has not approved a vaccine for Johne's disease — there is one used with good results in Norway — and there is no vaccine yet for the AIDS-like caprine arthritis encephalitis (CAE). Nor are there effective vaccines yet developed for pneumonia or coccidiosis.

Remember that a vaccine is not medicine, in the sense of being a "cure." It's a preventive measure. If your goat has already come down with enterotoxemia, it's too late for a vaccination.

Also keep in mind that vaccines often differ according to their manufacturer. No book can give blanket instructions regarding these products, but even more importantly, just because you used a vaccine for a given purpose once, don't assume it will be the same next time. Always read and follow instructions on the labels carefully.

There are other vaccinations, one for soremouth being one of the more prominent. But many goat owners resist vaccinating kids for soremouth because it sometimes gives kids a bad case of the disease.

Of course, there are always dangers. All injections bring the risk of an anaphylactic reaction (anaphylaxis is sensitivity to drugs or foreign proteins introduced into the body resulting from sensitization following prior contact with the causative agent). Always watch your animals for 10 minutes after any injections, and have a bottle of epinephrine on hand, just in case. And keep your veterinarian's phone number handy.

A GOAT-KEEPING CALENDAR: PLOTTING TASKS FOR A YEAR

January

Set up your record keeping for the year. What information and how much depends on why you keep goats and how much you enjoy keeping (or looking back at) records. Basic data should include production and health records to aid in culling. Include dates of deworming and vaccinations and products used. Record expenses; knowing what your milk costs is a great incentive and tool to improve your management. Record birth dates, weights, and other pertinent information, including sire and dam (no, you will not remember these a few years from now). If you raise registered animals or show, you will need to record much more.

Check your goats for lice and dust with a louse powder if necessary.

Vaccinate with CDT 45 to 60 days before kidding.

When you order seeds for your garden, order some for a "goat garden" too! Include carrots, kale, chard, collards, and comfrey roots.

Have everything ready for kidding, including iodine, and feed pans or bottles and nipples. You might not need a heat lamp, but have one on hand, just in case. An extra bulb is good insurance.

February

If any of your does were bred in September, they'll kid in February.

Frozen water buckets or troughs can mean extra work at this time of the year, but fresh water is important.

Disbud kids before they are 2 weeks old. Castrate buck kids by 4 weeks.

February is the "make-it-or-break-it reality month," according to Sheila Nixon, who has run a commercial dairy since 1958 and grew up on a goat dairy. The rush season of kidding and the worst weather of the year separates the dreamers from the real goat people, she says.

Spend time with your kids. Handling and talking to them when they're young will make them easier to handle when they grow up.

March

Does bred in October will freshen this month.

How are your February kids doing? Check their progress with a monthly weigh-in. For the first 5 months they should be gaining 10 pounds (4.5 kg) a month. In other words, if the birth weight was 8 pounds (3.5 kg), at 1 month of age the kid should weigh 18 pounds (8 kg); 2 months, 28 pounds (13 kg). (If you didn't weigh them at birth, figure 8 pounds as an average birth weight.) If they gain faster, good; slower, check your management.

Check your pasture fences, and plan for any needed repairs.

If rotational grazing will work for you, lay out your paddocks now. Build the fences or order movable fencing.

Is there an Easter or Passover market for your extra kids?

Daylight saving time starts on the second Sunday of March. If you're the punctual type, you can "spring ahead" on chores time about 10 minutes a day for 6 days to ease yourself and your goats into the new schedule.

Magnesium deficiency and grass tetany can occur from early spring grazing on lush pasture, which may be high in potassium. To prevent this, feed hay first, and limit the time spent on pasture.

Thoroughly clean pens and stalls when weather permits.

April

Are there muddy spots in the goat yard? Fill, or drainage, may be called for. Are there any plants in your goat yard or pasture that might be harmful to your animals? Borrow or purchase a book written for your locale to identify plants you're not familiar with. Obtain a list of dangerous plants from your Extension office.

Harvest dandelion greens as a feed supplement.

As your goats move onto pasture, watch for bloat. Feeding dry hay before letting the goats onto pasture for the day is a good preventive practice.

May

Control flies before they become a problem. The number-one defense is sanitation: don't give them a place to get started. Keep bedding clean and dry, and don't let wet spots accumulate around watering devices or buckets. You might also want to use parasitic wasps, diatomaceous earth, flypaper, and traps, in any combination. Chemical sprays should be a last resort in a home dairy.

As the weather warms, scrub water buckets and troughs more frequently to eliminate scum and algae.

Clip goats to keep them cleaner and cooler and to discourage external parasites. While you're there, trim hooves.

June

Time to make hay or to buy it "from the field" while neighbors are baling, so you get the best price. To ensure quality and a fair price, have the hay tested (see your county Extension agent for details).

Milk production is peaking. Use the surplus to make cheese, but also freeze some for the winter drought. If you still have too much, feed it to a calf, a pig, or chickens or cull your herd.

If you see a goat that's trembling, is breathing rapidly and shallowly, and has a rising body temperature, it's probably heat exhaustion. Provide plenty of clean, cool water, electrolytes, and trace minerals. Provide shade to prevent the problem in the future.

Goats can become sunburned; white-skinned animals are most susceptible. Some people say that eating lush clover or buckwheat increases the chances for sunburn.

July

Hot weather means goats need more water and plenty of shade. If animals are to be moved, do it at night or on cool days.

Rotate pastures as needed for optimum forage production and herd nutrition and health. Mow the weeds the goats won't eat before they go to seed, to encourage more palatable forage.

By now, February kids should weigh 50 pounds (23 kg).

Be aware that rain after a long dry spell can increase the nitrate content of some common pasture plants, which could result in poisoning.

Pinkeye becomes more common in hot, dry weather. Watch your goats to be sure their eyes don't water excessively or cloud over.

Now is the time to make goat-milk ice cream!

You might try flushing, or temporarily increasing energy and protein in feeds, to stimulate estrus, synchronize breeding, and increase litter size.

August

This is the time to deworm and to give any needed vaccinations.

Check your production records. Decide which does will be bred early or late or might be milked through and which should be culled to reduce the winter feed bills.

Breed by weight, not by age. A doeling bred when she has achieved about half her projected adult weight will be more productive, efficient, economical, and healthy than one bred later. This weight is usually around 80 pounds (35 kg).

As the days get shorter, your does could start cycling, especially if there is a cold weather snap. Record heat periods on your calendar. Prepare for a trip to the buck 16 to 18 days later for early breeding, or just keep track of the heat periods for later reference. (If you don't own a buck, line one up now.)

Whether you plan to keep or sell doe kids, find the right buck for each doe, one that will improve your herd.

Feed the goats (buck and does) carrots for vitamin A, good greens, and a small amount of oats and bran, but cut down on legumes, which some say can affect fertility.

Be sure all fences, gates, and latches are in good condition and sturdy enough to prevent breeding accidents.

September

How about a goat barbecue for Labor Day?

As does are bred, mark the date on your calendar as well as the expected birth date (150 days later). A feed high in fiber and lower in protein than the milking ration is in order for the first 3 months of gestation for a doeling. Your milking doe should be fed the milking ration, but aim for 1 pound (0.45 kg) of grain to 2 to 3 pounds (1 to 1.5 kg) of milk produced.

Check your feed supply, and estimate your needs for the winter. Do you have enough hay and grain on hand? Bedding? If you lack storage space, have you locked in a regular and reliable source of quality feed?

Seed or reseed pastures.

October

Be sure your goats' housing is draft-free but well ventilated. Autumn's rapid weather changes increase the potential for pneumonia. Protect your goats from drafts, and keep them dry.

Be sure your goats don't overindulge on apples or other fall produce that's available now. Feed these with care, and watch for bloat.

Frost can change the chemical composition of many forage plants, including johnsongrass, sorghum, Sudan, and alfalfa, making them toxic.

November

When you rake leaves, bag them for winter goat treats (but not wild cherry leaves!). Or if you have large quantities, use them as bedding. They will make even better compost after being in the goat barn all winter. Or use them to mulch the carrots heavily so they can be dug during the winter as a special treat — for you and the goats.

Daylight saving time ends (first Sunday in November). As in the spring, you may want to gradually shift your milking schedule.

Be sure your goats get enough exercise to avoid pregnancy toxemia.

Take steps to prevent water lines from freezing.

Check again for drafty conditions in the barn.

December

Reduce the amount of grain fed to does 4 months after breeding.

Are any does still cycling? Time is running out. Pen breeding might be in order (let her run with the buck).

Inventory feed and bedding again.

Check kidding supplies.

Check and trim hooves. They grow faster when the goats are on soft bedding than when they're in rocky pastures.

Do not feed a discarded Christmas tree to the goats. Some trees are treated with chemicals, and too much of a good thing may cause abortions in does (see page 124).

MAKING AND FEEDING SILAGE

Occasionally, a *Countryside* reader asks about feeding silage to goats. Yes, they will eat it, and yes, you can make it in plastic garbage bags, even from lawn clippings. But it's much more complex than that, and very few goat owners bother because, for most, it's more trouble than it's worth. Silage also increases the risk of listeriosis — a deadly neurologic disease — from dirt contamination. Still, it doesn't hurt to know the basics so you can judge for yourself.

A silo is a huge pickling vat. Think of a sauerkraut crock 12 feet (3.5 m) or more in diameter and often 80 feet (24 m) high or more. Chopped corn — the whole plant: cobs, stalks, and all (corn silage) — or chopped hay (haylage) is blown into the top of the silo through a pipe, much like insulation is blown into a house. Proper chopping is essential to ensure good packing, which eliminates air. It is firmly packed and left to ferment, just like kraut, for about 21 days. Note that the moisture content is of extreme importance. If it's too dry, it won't ferment; too wet, and it will rot.

Silage, or haylage, is a standard feed on most cow dairy farms because it's nutritious, easy to handle, and very cheap compared to hay and grain. But these attributes are hard to capture in a goat dairy, especially a small one. One of the primary reasons is spoilage, which occurs when silage is exposed to air, so a certain amount must be used every day. With a conventional silo, this is easy for cow herds but almost impossible for most goat herds.

Farmers like corn silage because it utilizes the entire plant, thus producing more feed (and milk) per acre. In northern areas, this use can salvage corn that doesn't mature enough to harvest as grain.

Haylage doesn't require the dry weather needed to make baled hay and can even allow you to salvage forages that were intended for hay but got caught by wet weather. It can also be made early or late in the season, when good drying weather is often scarce. Haylage can be made from alfalfa or other legumes; forages such as fescue and other grasses; or cereal grains in the "boot" stage, including oats and rye.

In one Midwestern study, alfalfa silage had 4 percent more protein than alfalfa of the same quality put up as hay, because of reduced leaf loss. In addition, the dry matter harvest was 27 percent higher.

Silage is much more mechanized than either grain or baled hay. There are no bales to store in the haymow and no bales to throw down at feeding time. No grain dryers, no grinding and mixing. An auger on a silo unloader scrapes off the top layer and shoots it down to a waiting cart (often electric, these days) that delivers it to the animals. In some operations it goes directly to the feed bunks.

Feed Value

As for feed value, the fermentation in the absence of air causes the sugars to break down, which is like predigesting the feed for the animals. The acidity increases, which accounts for its keeping qualities, but this keeping quality is lost in the presence of air. Oxygen activates the bacteria, and the silage heats up and molds within 2 to 3 days. That, and the cost of equipment, is the bad news for goat milkers.

If enough material can be removed from the top of the silage every day, this is no

GARBAGE BAG SILAGE

Despite its problems, silage appeals to many small landholders, especially when they hear that it can be made from lawn clippings, in garbage bags! The method differs from "real" silage only in scale. The requirements for moisture levels and air exclusion are the same. But remember, if grass has been treated with chemical fertilizers or pesticides, it's not safe for goats. Here's how it's done.

1. Mow the grass, and let it dry but not too much. Depending on the weather and humidity and the moisture content of the grass, drying might take only a few hours or it might take a day or more. Remember: too dry, and it won't ferment; too wet, and it will rot (good haylage has 30 to 35 percent dry matter, but more is recommended for bag silage).

2. Rake up the grass, and eliminate any twigs or stems that might puncture the bag. Be very careful to not pick up dirt in the grass. Pack the clippings tightly into the strongest plastic bags you can find. Sit on the bag to force out as much air as you possibly can. Tie the top tightly, using baling twine, not a twist tie.

3. Store the bags out of the sun for 3 weeks. As long as the bags don't get punctured or opened, the silage will keep for many months.

Note: When you open the bag, there might be a thin layer of white mold on the top. This is normal, since it's almost impossible to exclude all the air, which activates the bacteria in the grass. Put the moldy stuff in the compost bin. If the whole bag of grass is a slimy, stinking mess, it's all compost. Either the grass went into the bag too wet, or there was air in the bag. Good grass silage is green and has a fermented but pleasant aroma. If you're really short of feed, or money to buy it, bag silage might be worth the effort. Under any circumstances it could be fun to try it, just as an experiment (but the average backyard goat dairy will be ahead in time and money by feeding purchased hay and a prepared goat grain ration).

Use a bag of silage, once opened, within 2 to 3 days. If you have just a few goats, they won't eat this much, especially at first when you'll naturally want to limit feeding it to avoid digestive upsets. Always introduce any new feed slowly, over a week or more.

"Garbage bag silage" can be made from lawn clippings, if you don't use pesticides on your lawn. Eliminate as much air as possible from the bag before closing it tightly, and be sure the plastic won't be punctured. Air will cause the clippings to mold and rot.

problem. But in even a small silo, this amounts to far more than most goat herds can consume in a day. With just a few animals, forget it.

There are other problems. We know of one case where a cow dairyman also had a number of goats. The cows supposedly took care of the spoilage problem, so the goats got silage, too. Some of them got sick. The problem was identified as listeriosis, traced to the silage. It didn't bother the cows.

Silage is a relatively recent agricultural development. Its original purpose was to replace labor-intensive root crops such as carrots, mangel beets, and turnips with more "industrialized" capital-intensive feeds. As farms got bigger and more mechanized, research concentrated on silage rather than root crops. Today, silage is common, while almost no one grows roots for livestock feed, though some homesteaders still do.

GESTATION CHART

This table is based on an average gestation of 150 days. Remember that goats don't read tables, so use the expected kidding date as an approximation. When the gestation includes February, an allowance is made for leap years in column 3.

Example: If your doe is bred September 22, she can be expected to kid February 19. If she is bred November 10, she can be expected to kid April 9 in most years and April 8 in a leap year.

A doe bred in:	... will kid on the same date in this month, minus the number in column 3	
September	February	−3
October	March	−1 or −2
November	April	−1 or −2
December	May	−1 or −2
January	June	−1 or −2
February	July	0 or −1
March	August	−3
April	September	−3
May	October	−3
June	November	−3
July	December	−3
August	January	−3

Dairy Goat Registries and Breed Organizations

Most states have regional or state-wide dairy goat associations that can be found on the Internet or through your local Extension office.

Alpines International Club
www.alpinesinternationalclub.com

American Dairy Goat Association
Spindale, North Carolina
828-286-3801
www.adga.org

American Goat Society
Pipe Creek, Texas
830-535-4247
www.americangoatsociety.com

American LaMancha Club
www.lamanchas.com

International Nubian Breeders Association
www.i-n-b-a.org

International Sable Breeder's Association
http://sabledairygoats.com

National Saanen Breeders Association
http://nationalsaanenbreeders.com

National Toggenburg Club
http://nationaltoggclub.org

Nigerian Dwarf Goat Association
www.ndga.org

Oberhasli Breeders of America
oberhasli.webs.com

Vaccines and Medicines

Nasco
Fort Atkinson, Wisconsin
800-558-9595
www.enasco.com/farmandranch

PBS Animal Health
Massillon, Ohio
800-321-0235
www.pbsanimalhealth.com

Premier1 Supplies
Washington, Iowa
800-282-6631
www.premier1supplies.com

Sydell, Inc.
Burbank, South Dakota
800-842-1369
www.sydell.com

Goat Supplies

There are other livestock companies that include a page or two of goat equipment, but these are the ones that have a wide range of specialty supplies.

Caprine Supply
De Soto, Kansas
800-646-7736
www.caprinesupply.com

Hamby Dairy Supply
Maysville, Missouri
816-449-1314
http://hambydairysupply.com

Jeffers
Dothan, Alabama
800-533-3377
www.jefferspet.com

Cheesemaking, Sausage, and Soapmaking Supplies

Bramble Berry Soap Making Supplies
Bellingham, Washington
360-676-1030
www.brambleberry.com

The Cheese Maker
Mequon, Wisconsin
414-745-5483
www.thecheesemaker.com

Eldon's Sausage and Jerky Supply
Spokane Valley, Washington
800-352-9453
www.eldonsausage.com

The Essential Oil Company
Portland, Oregon
800-729-5912
www.essentialoil.com

Homesteader's Supply Inc.
Sparta, Tennessee
928-583-0254
www.homesteadersupply.com

Leeners
330-467-9870
www.leeners.com

Mold Market
509-245-3620
www.moldmarket.com

New England Cheesemaking Supply Company
South Deerfield, MA
413-397-2012
www.cheesemaking.com

The Sausage Maker, Inc.
Buffalo, New York
888-490-8525
www.sausagemaker.com

Sweet Cakes Soapmaking Supplies
Minnetonka, Minnesota
952-945-9900
www.sweetcakes.com

Recommended Reading

State Extension publications (research-based information specific to your area). Contact your county Extension agent or find bulletins online.

Boldrick, Lorrie, and Lydia Hale. *Pygmy Goats: Management and Veterinary Care, 4th ed.* All Publishing, 2009.

Carroll, Ricki. *Home Cheese Making, 3rd ed.* Storey Publishing, 2002.

Cavitch, Susan Miller. *The Soapmaker's Companion*. Storey Publishing, 1997.

Mikela, Casey. *Milk-Based Soaps*. Storey Publishing, 1997.

Peery, Susan Mahnke, and Charles G. Reavis. *Home Sausage Making*, 3rd ed. Storey Publishing, 2003.

Smith, Mary C., and David M. Sherman, *Goat Medicine*, 2nd ed. Wiley-Blackwell, 2009.

Toth, Mary Jane. *Goats Produce Too! The Udder Real Thing*. Vol. 2, *Cheese Making & more*. Mary Jane Toth, 1998.

Watson, Anne L. *Milk Soapmaking*. Shephard Publications, 2009.

Periodicals

Dairy Goat Journal
800-551-5691
www.countrysidenetwork.com/magazines/dairy-goat-journal/

United Caprine News
817-297-3411
www.unitedcaprinenews.com

INTERIOR PHOTOGRAPHY BY © 5PH/iStockphoto.com, 16; © Andrew Cline/Alamy Stock Photo, 74; © Anne Lewis/Alamy Stock Photo, 14 bottom; © Arterra/Getty Images, 83, 88 (dandelion); © asiseeit/iStockphoto.com, 205; © ATTA KENARE/Getty Images, 3; © bamboofox/Alamy Stock Photo, 218; © blickwinkel/Alamy Stock Photo, 89 (wild cherry); © Boston Globe/Getty Images, 225; © Catalin Petolea/Alamy Stock Photo, 152; © Cheryl Fleishman/Alamy Stock Photo, 32; © Cultura Creative (RF)/Alamy Stock Photo, 237; © Daniel Lang/Alamy Stock Photo, 14 top; © Dasha Rosato/Alamy Stock Photo, 145; © David Page/Alamy Stock Photo, 165; © Denny Ellis/Alamy Stock Photo, 186; © Design Pics Inc./Alamy Stock Photo, 251; © Diane Randell/Alamy Stock Photo, 47; © Duncan Soar/Alamy Stock Photo, 249; © Edwin Remsberg/Alamy Stock Photo, 154; © essphotographer/iStockphoto.com, 66; © eyetoeyePIX/iStockphoto.com, 181; © Florapix/Alamy Stock Photo, 89 (hemlock); © flowerphotos/Alamy Stock Photo, 89 (mountain laurel); © Frederic Didillon/Getty Images, 90 left; © gerenme/iStockphoto.com, 26; © H. Mark Weidman Photography/Alamy Stock Photo, 89 (locoweed); © Ian Rose/Getty Images, 103; © imageBROKER/Alamy Stock Photo, 30, 89 (dock, sorrel); © Inga Spence/Alamy Stock Photo, 269; © Isaac74/iStockphoto.com, 18; © itsabreeze photography/Getty Images, 88 (daisy); © Janet Horton/Alamy Stock Photo, 9, 48, 51, 58; © JannHuizenga/iStockphoto.com, 76; © Jason Houston, ii, vii, 172, 180, 185, 226; © John Polak, 222; © kreinick/iStockphoto.com, 241; © Lee O'Dell/123RF, 79; © Lena Ason/Alamy Stock Photo, 175; © Les Gibbon/Alamy Stock Photo, 212; © Lynn Stone, 5, 6, 10–12, 75, 104, 142, 146, 149, 190, 200, 216; © mackoflower/123RF, 70; © Marianne Mayer/123RF, 162; Mars Vilaubi, 85, 88 (nettle, yarrow), 89 (rhubarb), 95, 106, 108 right, 116, 120, 208; © Mike Powles/Getty Images, 113; © Mint Images Limited/Alamy Stock Photo, viii; © My Lit'l Eye/Alamy Stock Photo, 68–69; © Naoya Maruo/Getty Images, 88 (multiflora rose); © Nigel Cattlin/Alamy Stock Photo, 90 right, 160; © Nigel Cattlin/Visuals Unlimited, Inc./Getty Images, 88 (thistle); © Peter Steiner/Alamy Stock Photo, 89 (milkweed); © Philip Game/Alamy Stock Photo, 71; © Photo Alto/Alamy Stock Photo, 60–61; © PhotoCuisine RM/Alamy Stock Photo, 55; © PicturePartners/iStockphoto.com, 232; © Ramn Gausachs/Getty Images, 88 (chicory); © Ross Drinkwater/Alamy Stock Photo, 89 (bracken fern); © sasimoto/iStockphoto.com, 238; © Sataporn Jiwjalaen/123RF, 108 left & center; © SCOTT CAMAZINE/Getty Images, 88 (plantain); © Scott W. Smith/agefotostock.com, 7; © Terryfic3D/iStockphoto.com, 89 (oak leaf); © Visuals Unlimited, Inc./Wally Eberhart/Getty Images, 91; © *The Washington Post*/Getty Images, 33; © xalanx/123RF, 178; © ZUMA Press, Inc./Alamy Stock Photo, 123

abomasum. The fourth or true stomach of a ruminant where enzymatic digestion occurs.

abscess. Boil; a localized collection of pus.

ADGA. American Dairy Goat Association, the oldest and largest dairy goat registry in the United States.

afterbirth. The placenta and associated membranes expelled from the uterus after kidding.

AGS. American Goat Society, a registry.

AI. Artificial insemination.

alveoli. Tiny hollow spheres in the udder whose cells secrete milk. Singular: alveolus.

American. A doe that is seven-eighths purebred and recorded with ADGA; a buck that is fifteen-sixteenths purebred and recorded with ADGA.

anthelmintic. A drug that kills worms.

antitrypsin factor. A substance that prevents the enzyme trypsin in pancreatic juice from helping to break down proteins. Present in soybeans.

AR (advanced registry). A designation for a goat that has produced at least 1,500 pounds of milk in a 305-day lactation.

ash. The mineral matter of a feed; what is left after complete incineration of the organic matter.

balling gun. Device used to administer a bolus (a large pill).

barn records. A tally of daily milk production kept by the goat owner rather than by an official testing organization.

blind teat. A teat that does not allow passage of milk; blind can also refer to a nonfunctioning half of an udder.

bloat. An excessive accumulation of gas in the rumen and reticulum, resulting in distension.

bolus. A large pill for animals; also, regurgitated food that has been chewed (cud).

breed. Animals with similar characteristics of conformation and color, which when mated together produce offspring with the same characteristics; the mating of animals.

breeding season. The period when goats will breed, usually from September to December.

browse. Bushy or woody plants; to eat such plants.

buck. A male goat.

buck rag. A cloth rubbed on a buck and imbued with his odor and kept in a closed container; used by exposing to a doe and observing her reaction to help determine if she's in heat.

buckling. A young male.

Burdizzo. A castrating device that crushes the spermatic cords to render a buck or buckling sterile.

butterfat. The natural fat in milk; cream.

CAE (caprine arthritis encephalitis). A serious and widespread type of arthritis, caused by a retrovirus.

California mastitis test (CMT). A do-it-yourself kit to determine if a doe has mastitis.

caprine. Pertaining to or derived from a goat.

carbonaceous hay. Any hay that is not a legume (such as the clovers and alfalfa) including timothy, brome, johnsongrass, and Bermuda grass.

chevon. Goat meat.

cistern. Final temporary storage area of milk in the udder.

classification. A system of scoring goats based on appearance.

colostrum. The first thick, yellowish milk a goat produces after giving birth, rich in antibodies without which the newborn has little chance of survival.

concentrate. The nonforage part of a goat's diet, principally grain, but including oil meal and other feed supplements, that is high in energy and low in crude fiber.

confinement feeding. Feeding goats restricted to a barn and exercise yard; that is, nonpastured goats.

conformation. The overall physical attributes of an animal; its shape and design.

creep feeder. An enclosed feeder for supplementing the ration of kids, but which excludes larger animals.

cull. To remove a substandard animal from a herd; also, such a substandard animal.

dairy cleaning agents. Alkaline or acid detergents for washing milking equipment; iodine or chlorine compounds for sanitizing milking equipment.

dam. Female parent.

DHIA (Dairy Herd Improvement Association). A national association with regional affiliates that test and record milk production of cows and goats to inform feeding, breeding, and culling decisions.

DHIR (Dairy Herd Improvement Registry). A milk production testing program administered by dairy goat registries in cooperation with DHIA.

disbudding iron. A tool, usually electric, that is heated to burn the horn buds from young animals to prevent horn growth.

dished face. The concave nose of the Saanen.

doe. A female goat.

doeling. A young female.

drenching. Giving medication from a bottle.

dry period. The time when a goat is not producing milk.

drylot. An animal enclosure having no vegetation.

elastrator rings. Castrating rings resembling rubber bands; they are applied with a special tool called an elastrator to the scrotum so it will atrophy and fall off.

electrolyte. Mineral salts necessary for life, including sodium, potassium, calcium, and magnesium, and that are lost when a body loses more fluid than it can take in.

enterotoxemia. A bacterial infection, usually resulting in death; also called pulpy kidney disease and overeating disease.

FAMACHA. A method to identify anemia and likelihood of internal parasites by color of the goat's inner-eye.

feed additive. Anything added to a feed, including preservatives, growth promotants, and medications.

flushing. Feeding females more generously 2 to 3 weeks before breeding in order to stimulate the onset of heat, induce the shedding of more eggs, resulting in more offspring and improving the chances of conception.

forage. The hay or grassy portion of a goat's diet.

free choice. Free to eat at will with food (especially hay) always present.

freshen. To give birth (kid) and come into milk.

gestation. The time between breeding and kidding (average 150 days).

grade. A goat that is not purebred, or cannot be proven pure by registry records; any goat of mixed or unknown ancestry.

Grade A. A category of licensed dairy meeting strict regulations for equipment, milk handling, and sanitation.

green forage. The green, growing plant component of a goat's diet.

growthy. Description of an animal that is large and well developed for its age.

hand feeding. Providing a measured amount of feed at set intervals.

hand mating. Controlled breeding, as opposed to letting a male run loose with or in a pen of unbred females.

hay. Dried forage.

haylage. Silage made from hay plants such as alfalfa.

heat. Estrus; the condition of a doe's being ready to breed.

hermaphrodite. A sterile animal with reproductive organs of both sexes, generally associated with the mating of two naturally polled animals.

homozygous. Containing either but not both members of a pair of alleles.

hormone. A chemical secreted into the bloodstream by an endocrine gland, bringing about a physiological response in another part of the body.

horn buds. Small bumps from which horns grow.

IM (intramuscular). Within the muscle.

inbreeding. The mating of closely related individuals.

intradermal. Into or between the layers of the skin.

intraperitoneal. Within the peritoneal cavity.

intravenous. Within a vein.

IU (international unit). A standard unit of potency of a biologic agent such as a vitamin or antibiotic.

Johne's disease. A wasting, often fatal form of enteritis.

ketosis. Overaccumulation of ketones in the body, responsible for pregnancy disease, acetonemia, twin lambing disease, and others that occur at the end of pregnancy or within a month of kidding.

kid. A goat under 1 year of age; to give birth.

koumiss. A fermented goat-milk drink originally from central Asia and made of mare's milk. Also spelled *kumiss*.

lactation. The period in which a goat is producing milk; the secretion or formulation of milk.

lactation curve. Daily milk production as represented on a graph, usually rapidly rising soon after freshening, and then slowly falling.

legume. A family of plants having nodules on the roots bearing nitrogen-fixing bacteria, including alfalfa and the clovers.

linear appraisal. A system of scoring goats on individual conformation traits.

linebreeding. A form of inbreeding that attempts to concentrate the genetic makeup of some ancestor.

mastitis. Inflammation of the udder, usually caused by an infection.

microorganism. Any living creature of microscopic size, especially bacteria and protozoa.

milking bench (or stand). A raised platform, usually with a seat for the milker and a stanchion for the goat's neck, that a goat stands upon to be milked.

milking through. Milking a goat for more than 1 year.

milkstone. Cloudy, bacteria-inhabited film left by alkaline detergents.

New Zealand fencing. A system of electric fencing using a high-energy charger.

off feed. Not eating as much as normal.

out of. Mothered by.

overconditioned. Overfed; fat.

papers. Certificates of registration or recordation.

pedigree. A paper showing an animal's forebears.

pessary. A vaginal suppository, used after kidding to prevent infection if human assistance in the birth has been required.

polled. Naturally hornless.

precocious milker. A goat that produces milk without being bred.

protein supplement. A feed product containing more than 20 percent protein.

purebred. An animal whose ancestry can be traced back to the establishment of a breed through the records of a registry association.

raw milk. Milk as it comes from the goat; unpasteurized milk.

recorded grade. A goat, either not purebred or not verifiably purebred, that is recorded with ADGA.

registered. A goat whose birth and ancestry is recorded by a registry association.

rennet. An enzyme used to curdle milk and make cheese.

retained placenta. A placenta not expelled at kidding or shortly thereafter.

reticulum. The second compartment of the ruminant stomach.

rotational grazing. A system for pasturing livestock by which animals are turned out on one small section of pasture at a time; prevents overgrazing and sustains and renews plant growth.

roughage. High-fiber, low total digestible nutrient feed, consisting of coarse and bulky plants or plant parts; dry or green feed with over 18 percent crude fiber.

rumen. The first large compartment of the stomach of a goat where cellulose is broken down.

scours. Persistent diarrhea in young animals.

service. Mating.

settled. Having become pregnant.

silage. Fodder preserved by fermentation; also called haylage.

sire. Male parent; to father.

soiling. Harvesting and bringing feed to goats.

stanchion. A device for restraining a goat by the neck for feeding or milking.

standing heat. The period during which a doe will accept a buck for mating, usually about 24 hours.

star milker. A designation of high milk production based on a 1-day test, not the entire lactation. ★M, ★ ★ M, and so on indicates that the dam and granddam also held ★M status. ★B or star buck indicates star milkers in a buck's family tree.

straw. Dried plant matter (usually oat, wheat, or barley leaves and stems) used as bedding; also, the glass tube semen is stored in for AI.

streak canal. Opening at the end of a teat, surrounded by sphincter muscles.

strip. To remove the last milk from the udder.

strip cup. A cup into which the first squirt of milk from each teat is directed and which will show any abnormalities that might be in the milk.

subcutaneous. Beneath the skin.

synthesis. The bringing together of two or more substances to form a new material.

tattoo. Permanent identification of animals produced by placing indelible ink under the skin, generally in the ear but in the tail web of La Manchas.

test (to be on test; official test). To have daily milk production weighed and its butterfat content determined by a person other than the goat's owner.

therm. Unit of measurement of energy, used with animal feeds instead of calories. One therm is the amount of heat required to raise the temperature of 1,000 kilograms of water 1°C (1 therm = 1,000,000 calories).

total digestible nutrient (TDN). The energy value of a feed.

toxic. Of a poisonous nature.

trace mineral. A mineral nutrient essential to animal health but used only in very minute quantities.

type. The combination of characteristics that makes an animal suited for a specific purpose, such as "dairy type" or "meat type."

udder. An encased group of mammary glands provided with a teat or nipple.

udder wash. A dilute chemical solution, usually an iodine compound, for washing udders before milking.

unrecorded grade. A grade goat not recorded with any registry association.

upgrade. To improve the next generation by breeding a doe to a superior buck.

vermifuge. Any chemical substance administered to an animal to kill internal parasitic worms.

wattle. Small, fleshy appendage. Wattles are hereditary, not all goats have them, and they serve no useful purpose.

wether. A castrated buck.

whey. The liquid remaining when the curd is removed from curdled milk when making cheese.

WMT (Wisconsin mastitis test). A do-it-yourself kit to determine if a doe has mastitis.

METRIC CONVERSION CHARTS

VOLUME EQUIVALENTS

US	Metric
1 teaspoon	5 milliliters
1 tablespoon	15 milliliters
¼ cup	60 milliliters
½ cup	120 milliliters
1 cup	240 milliliters
1¼ cups	300 milliliters
1½ cups	350 milliliters
2 cups	480 milliliters
2½ cups	600 milliliters
3 cups	700 milliliters

TEMPERATURE CONVERSION

To convert Fahrenheit to Celsius, subtract 32 from Fahrenheit temperature, multiply by 5, then divide by 9.

Easy-to-Remember Equivalents

0°C = 32°F	30°C = 86°F
10°C = 50°F	40°C = 104°F
20°C = 68°F	Every 10°C = 18°F

VOLUME CONVERSION

To convert	to	multiply
teaspoons	milliliters	teaspoons by 4.93
tablespoons	milliliters	tablespoons by 14.79
fluid ounces	milliliters	fluid ounces by 29.57
cups	milliliters	cups by 236.59
cups	liters	cups by 0.24
pints	milliliters	pints by 473.18
pints	liters	pints by 0.473
quarts	milliliters	quarts by 946.36
quarts	liters	quarts by 0.946
gallons	liters	gallons by 3.785

WEIGHT EQUIVALENTS

US	Metric
0.035 ounce	1 gram
¼ ounce	7 grams
½ ounce	14 grams
1 ounce	28 grams
1¼ ounces	35 grams
1½ ounces	43 grams
1¾ ounces	50 grams
2½ ounces	70 grams
3½ ounces	100 grams
4 ounces	113 grams
5 ounces	140 grams
8 ounces	227 grams
8¾ ounces	250 grams
10 ounces	284 grams
15 ounces	425 grams
16 ounces (1 pound)	454 grams

LENGTH / AREA CONVERSION

To convert	to	multiply by
inches	centimeters	2.54
feet	meters	0.31
yards	meters	0.91
miles	kilometers	1.61
square feet	square meters	0.09
acres	hectares	0.41

WEIGHT CONVERSION

To convert	to	multiply
ounces	grams	ounces by 28.35
pounds	grams	pounds by 453.6
pounds	kilograms	pounds by 0.45

Page numbers in *italics* indicate drawings and photographs.
Page numbers in **bold** indicate tables and charts.

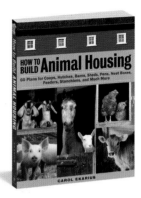

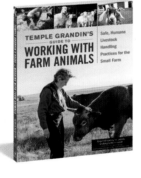

STOREY'S GUIDE TO RAISING

The Definitive Series for Essential Animal Husbandry Information

MORE THAN 2.2 MILLION COPIES SOLD!

This best-selling series offers fledgling farmers and seasoned veterans alike what they most need to know to ensure both healthy livestock and profits. Each book includes information on selection, housing, space requirements, behavior, breeding and birthing, feeding, health concerns, and remedies for illnesses. They also cover business considerations and marketing products that come from the animals.

THE COMPLETE STOREY'S GUIDE TO RAISING LIBRARY INCLUDES:

Beef Cattle
by Heather Smith Thomas

Chickens
by Gail Damerow

Dairy Goats
by Jerry Belanger and Sara Thomson Bredesen

Ducks
by Dave Holderread

Horses
by Heather Smith Thomas

Keeping Honey Bees
by Malcolm T. Sanford and Richard E. Bonney

Meat Goats
by Maggie Sayer

Miniature Livestock
by Sue Weaver

Pigs
by Kelly Klober

Poultry
by Glenn Drowns

Rabbits
by Bob Bennett

Sheep
by Paula Simmons and Carol Ekarius

Training Horses
by Heather Smith Thomas

Turkeys by Don Schrider